T0328372

Multi-Robot Exploration
for Environmental Monitoring

Multi-Robot Exploration for Environmental Monitoring

The Resource Constrained Perspective

Kshitij Tiwari
Dept. of Electrical Engg. & Automation (EEA)
Aalto University
Espoo, Finland

Nak Young Chong
School of Information Science
Japan Advanced Institute of Sci. & Tech.
Nomi City, Ishikawa, Japan

ACADEMIC PRESS
An imprint of Elsevier

Academic Press is an imprint of Elsevier
125 London Wall, London EC2Y 5AS, United Kingdom
525 B Street, Suite 1650, San Diego, CA 92101, United States
50 Hampshire Street, 5th Floor, Cambridge, MA 02139, United States
The Boulevard, Langford Lane, Kidlington, Oxford OX5 1GB, United Kingdom

Notices

Knowledge and best practice in this field are constantly changing. As new research and experience
broaden our understanding, changes in research methods, professional practices, or medical treatment
may become necessary.

Practitioners and researchers must always rely on their own experience and knowledge in evaluating and
using any information, methods, compounds, or experiments described herein. In using such
information or methods they should be mindful of their own safety and the safety of others, including
parties for whom they have a professional responsibility.

To the fullest extent of the law, neither the Publisher nor the authors, contributors, or editors, assume
any liability for any injury and/or damage to persons or property as a matter of products liability,
negligence or otherwise, or from any use or operation of any methods, products, instructions, or ideas
contained in the material herein.

Library of Congress Cataloging-in-Publication Data
A catalog record for this book is available from the Library of Congress

British Library Cataloguing-in-Publication Data
A catalogue record for this book is available from the British Library

ISBN: 978-0-12-817607-8

For information on all Academic Press publications
visit our website at https://www.elsevier.com/books-and-journals

Publisher: Mara Conner
Acquisition Editor: Sonnini R.Yura
Editorial Project Manager: Thomas Van Der Ploeg
Production Project Manager: R.Vijay Bharath
Designer: Miles Hitchen

Typeset by VTeX

Working together
to grow libraries in
developing countries

www.elsevier.com • www.bookaid.org

Dedication

In loving memory of my mother who managed to raise me and acquire the Post-graduate degree seamlessly. It was always your dream to see me make a mark for myself. Well, having obtained the Ph.D., this is the next step towards that dream, and I dedicate this book to you. Hope, the angels are taking good care of you and that you are resting in peace. Lots of love and respect.

Dr. Kshitij Tiwari

Dedicated to my beloved family for their love and sacrifice.

Prof. Nak Young Chong

Preface

This book describes an efficient environment monitoring application assisted by mobile robots. Traditionally, environmental monitoring was carried out using static sensors positioned across the field of interest but obtaining dense spatial resolution in this case required many sensors to be placed. This is where robots come to aid given their agility and mobile sensing capabilities. This research has a dual focus: *firstly*, the computational resources of the robots themselves, and *secondly*, the performance of the models being used to represent the environmental phenomenon.

Environmental monitoring basically entails inferring the spatially varying and temporally evolving dynamics of the target phenomenon. The possible applications could include pollution, algal bloom, oil spill, nuclear radiation spill monitoring, and other related domains. However, the area to be monitored is usually large and the environmental dynamics being modeled are generally complex.

Multi-Robot Exploration for Environmental Monitoring is intended to demonstrate the validity of an interdisciplinary approach that combines state-of-the-art machine learning models with robot navigation and exploration. The majority of this work was performed during my Ph.D. tenure at the Japan Advanced Institute of Science & Technology (JAIST), Japan. This book, however, is an adaptation of my original Ph.D. thesis. Through this book, I attempt to emphasize the need for accounting both robot (hardware) and machine learning (software) constraints when analyzing applied machine learning applications in robotic domains. With this, I introduce the novel *Resource Constrained Perspective* which is a one-of-a-kind book length discussion about balancing the trade-off between hardware to software when deploying robots for real-world applications. I believe that with this book several new doors may open to further this research and enhance the current findings.

While writing and publishing this book, I have taken care to keep this book as stand-alone and self-contained as possible. I have tried to include as much resources as are deemed necessary within the book, and for additional references I have pointed to the appropriate works. As my father always puts it, $Research = Re + Search$, i.e., *research* basically entails searching for what is already known and building upon the existing results. Thus, while I am grateful to all the authors and scholars before me for laying the foundation of this work, I do duly acknowledge their respective contributions throughout this book. Also, I would like to thank my then supervisor Prof. Nak Young Chong who is also the co-author of this book. It was him who laid the breadcrumbs that pointed me eventually in this direction.

Kshitij Tiwari
Aalto University, Espoo, Finland
July 10, 2019

About the authors

Kshitij Tiwari is a Postdoctoral Researcher at the Department of Electrical Engineering & Automation, Aalto University, Finland. He works with the Intelligent Robotics group within the Department. He received his Ph.D. degree in 2018 from the Japan Advanced Institute of Science & Technology (JAIST), Japan. He obtained the M.Sc. degree in Artificial Intelligence with a special focus in Intelligent Robotics from the University of Edinburgh in 2014 and the BEngg degree in Electronics & Communication from the University of Hong Kong in 2013. His research interests include (but are not limited to) field robotics, applied machine learning, bio-inspired SLAM, path planning under uncertainty, and related domains.

Nak Young Chong is a Professor in Robotics at JAIST, Japan. He received his BS, MS, and Ph.D. degrees in mechanical engineering from Hanyang University, Seoul, Korea, in 1987, 1989, and 1994, respectively. From 1994 to 2007, he was a member of research staff at Daewoo Heavy Industries and KIST in Korea, and MEL and AIST in Japan. In 2003, he joined the faculty of Japan Advanced Institute of Science and Technology (JAIST), where he currently is a Professor of Information Science. He also served as Vice Dean for Research and Director of the Center for Intelligent Robotics at JAIST. He was a Visiting Scholar at Northwestern University, Georgia Institute of Technology, University of Genoa, and Carnegie Mellon University, and also served as an Associate Graduate Faculty at the University of Nevada, Las Vegas, International Scholar at Kyung Hee University, and as Distinguished Invited Research Professor at Hanyang University. He serves as Senior Editor of the IEEE Robotics and Automation Letters and Intelligent Service Robotics, Topic Editor-in-Chief of International Journal of Advanced Robotic Systems, and served as Senior Editor of IEEE ICRA CEB, and IEEE CASE CEB, and as Associate Editor of the IEEE Transactions on Robotics and Intelligent Service Robotics. He served as Program Chair/Co-Chair for JCK Robotics 2009, ICAM 2010, IEEE Ro-Man 2011, IEEE CASE 2012, IEEE Ro-Man 2013, URAI 2013/2014, DARS 2014, ICCAS 2016, and IEEE ARM 2019. He was a General Co-Chair of URAI 2017 and is serving as General Chair of UR 2020. He also served as Co-Chair for IEEE-RAS Networked Robots Technical Committee from 2004 to 2006, and Fujitsu Scientific System Working Group from 2004 to 2008.

Acknowledgments

I would like to thank Prof. Nak Young Chong for helping me in editing and putting this book together. A special mention for my father, Dr. Anupam Tiwari, who has always been my guiding light through tough times and otherwise. Also, a special word of thanks to Valentin Honoré and Dang T.L. Quyen, who have been my support system through my Ph.D. days. I am glad to have known you guys and hope our friendship lasts forever.

We thank our editorial project manager, Thomas Van Der Ploeg & Isabella c. Silva for guidance throughout the publication process. We would also like to thank our production project manager, R. Vijay Bharath, acquisition Editor, Sonnini R. Yura, publisher, Mara Conner, our designer, Miles Hitchen and copyrights coordinator Narmatha Mohan for their respective contributions to this endeavour. We are also grateful to all the anonymous referees for carefully reviewing our proposal in the early stages of this project.

Kshitij Tiwari
July 10, 2019

Contents

List of figures

List of tables

Nomenclature

Acronyms

AE	Energy available for ancillary functions (J)
ME	Energy available for maneuvering functions (J)
ABC	Approximate Bayesian Computation
ADAM	Adaptive Moment Estimation
ARD	Automatic Relevance Determination
ARMA	Auto-Regressive Moving Average
ASV	Autonomous Surface Vehicle
AUV	Autonomous Underwater Vehicle
BO	Bayesian Optimization
BCM	Dirichlet Process
BGD	Batch Gradient Descent
BI	Bayesian Inference
CEO	Chief Executive/Estimation Officer
CPP	Coverage Path Planning
CSP	Coverage Salesman Problem
DAS	Decentralized Active Sensing
DDD	Dirty-Dangerous-and-Dull
DFO	Derivative Free Optimization
DPM	Dynamic Power Management
DPMGP	Dirichlet Process Mixture of GP
DVS	Dynamic Voltage Scaling
EV	Electric Vehicle
FOV	Field of View
FPS	Frames Per Second
FuDGE	Fusion of Distributed Gaussian Process Experts
GCM	General Circulation Model
GD	Gradient Descent
GDR	Gradient Descent with Random Restarts
GMM	Gaussian Mixture Model
GP	Gaussian Process
GP-LVM	Gaussian Process Latent Variable Model
gPoE	Generalized Product-of-Experts
GPPS	Gaussian Process Positioning System
GPR	Gaussian Process Regression
GTGP	Ground Truth GP
HAB	Harmful Algal Bloom
HSIC	Hilbert–Schmidt Independence Criterion
IEM	Intelligent Environment Monitoring
i.i.d.	Independent and Identically Distributed
IPP	Informative Path Planning
KLD	Kullback–Leibler Divergence
LWF	Locally Weighted Fusion
MARTA	Multilevel Autonomy Robot Telesupervision Architecture
MBD	Mini-batch Gradient Descent

MLE	Maximum Likelihood Estimation
MNC	Multi-National Company
MODIS	Moderate Resolution Imaging Spectroradiometer
MoE	Mixture-of-Experts
MOGP	Multi-Output GP
MR-GP	Map Reduce GP
MRS	Multi-Robot System
MVN	Multi-Variate Normal
NERC	Natural Environment Research Council
NN	Nearest Neighbor
NOAA	National Oceanic and Atmospheric Administration
NOSTILL-GP	NOn-stationary Space TIme variable Latent Length scale GP
OASIS	Ocean–Atmosphere Sensor Integration System
OGM	Occupancy Grid Map
OP	Orienteering Problem
ORangE	Operational Range Estimation
PDF	Probability Density Function
PEM	Persistent Environment Monitoring
PoE	Product-of-Experts
PTTE	Potentially Toxic Trace Element
PVI	Partitioned Variational Inference
RC-DAS	Resource Constrained Decentralized Active Sensing
RC-DAS[†]	RC-DAS + Homing
RMSE	Root Mean Squared Error
ROMS	Regional Ocean Modeling System
RSB	Robot Sensor Boat
RSS	Received Signal Strength
SAR	Search and Rescue
SELFE	Semi-implicit Eulerian–Lagrangian Finite Element
SGD	Stochastic Gradient Descent
SoC	State of Charge
SPM	Suspended Particulate Matter
STBC	Space-Time Block Code
TAOSF	Telesupervised Adaptive Ocean Sensor Fleet
TOF	Time of Flight
TSP	Traveling Salesman Problem
UAV	Unmanned Aerial Vehicle
UGV	Unmanned Ground Vehicle
UMV	Unmanned Marine Vehicle
UNEP	United Nations Environment Programme
US-EPA	US Environment Protection Agency
UUV	Unmanned Underwater Vehicle
VI	Variational Inference
VRP	Vehicle Routing Problem
WSN	Wireless Sensor Networks
xORangE	$x \in \{\text{offline, online}\}$ Operational Range Estimation

Constants

$\Gamma = 0.27$	Figure of Merit for UAV propellers
$\rho = 1.2041$	Density of air (kg/m^3)
$g = 9.8$	Gravitational acceleration (m/s^2)

Gaussian Process

| μ | Mean function of GP or mean vector of MVN pdf |

$\boldsymbol{\mu}^*$	$= \boldsymbol{\mu}(\mathbf{x}^*)$ Prior mean function over the unobserved inputs
$\boldsymbol{\mu}_{f\|D}$	Posterior mean of GP
μ_Q	Fused mean
$\boldsymbol{\theta}$	$= \{\sigma_s, l_s, \sigma_n\}$ Set of hyper-parameters
l_s	Spatial length scale
$\mathcal{K}$	$= \mathcal{K}(\mathbf{x}^-, \mathbf{x}^-)$ Auto-covariance function amongst all pairs of $\mathbf{x}^-$
$\mathcal{K}(\cdot, \cdot)$	Specific entry from covariance function
$\mathcal{K}^*$	$= \mathcal{K}(\mathbf{x}^*, \mathbf{x}^-)$ Cross-covariance function amongst observed and unobserved inputs
$\mathcal{K}^{**}$	$= \mathcal{K}(\mathbf{x}^*, \mathbf{x}^*)$ Auto-covariance function amongst unobserved inputs
$\mathcal{K}_{f\|D}$	Posterior covariance of GP
λ	Eigen value
$\mathcal{LL}$	Log-marginal likelihood
σ_Q	Fused variance
Σ	Covariance matrix of MVN pdf
Σ_m	$\triangleq \mathrm{diag}(l_{lat}^2, l_{long}^2)$ Diagonal length scale matrix
$\mathbf{x}^-$	List of observed inputs
$\mathbf{x}_m^-$	Multiple visited locations for the agent m
$\mathbf{x}_m^+$	Multiple candidate locations for the agent m
$\mathbf{y}^-$	List of observed targets
σ_s	Signal standard deviation
$\mathbf{x}^+$	List of candidate inputs
$\mathbf{x}^*$	List of unobserved inputs
σ_n	Noise standard deviation
f	Underlying function to be modeled via GP

Notation

$\downarrow$	Decrease or drop in a quantity
$\uparrow$	Increase in a quantity
$(\cdot)^{[t]}$	Time-step during mission
α	Weights for RC-DAS objectives
$\odot^{[0:end]}$	Value of the quantity $\odot$ from $t = 0$ until end of mission
$\odot^{[0:t]}$	Value of the quantity $\odot$ from $t = 0$ to $t = t$
$\odot^{[t:end]}$	Value of the quantity $\odot$ from $t = t$ until end of mission
η	Energy loss (%)
γ	Terrain gradient
$\hat{\mathbf{x}}_*$	Next-best-location as per $*$ objective function
$\hat{\mathbf{x}}_m$	Next-best-location for the agent m
$\mathbb{H}$	Entropy
x	Single (input) location
y	Single (target) observation
$\mathbf{x}$	Multiple (inputs) locations as vector
$\mathbf{y}$	Multiple (targets) observations as vector
$\mathbb{I}$	Identity matrix
I_{motor}	Current flowing through the motor
$\mathcal{N}$	Normal distribution
$\mathcal{O}(\cdot)$	Complexity
$\mathcal{R}$	Set of real numbers
Q	Probe (test) input for FuDGE
ω	Angular velocity of robot (rad/s)
Ω	Overall system efficiency
${}^r\Omega_{man}$	Net maneuvering efficiency of robot r (%)
$\overline{[*]}$	Average quantity $*$
τ_D	Drag torque (N-m)
$\langle \cdot \rangle$	n-tuple

θ	Terrain elevation (degrees)
$\widehat{[*]}$	Estimated quantity $*$
$\widetilde{E}$	Reduced energy available from battery (J)
$\zeta(\cdot)$	Responsibility of GMM component
A	True area over which drag force is acting (m^2)
B	Budget allocated for exploration
B_{res}	Residual budget available for exploration
C_D	Zero drag coefficient
C_{rr}	Ground (rolling) resistance
con	Condition number
dom	Domain of target phenomenon
d	Travel distance (m)
d_{max}	Max distance (m)
D	$\triangleq 100 \times \left(\dfrac{T_M - T_A}{T_M} \right)$ Duty cycle (% of driving time) during which maneuvering energy is non-zero
T_A	Total stopping time for ancillary functions
T_M	Mission time
$\mathcal{E}$	Computational cost for informative path planning
E_{mech}	Supply energy that gets transformed into mechanical energy
E_{motor}	Energy supplied by batter to motors
$\widetilde{E}_{motor}$	Net energy for motors
E_O	Rated energy available from battery (J)
$\mathcal{F}$	Cost of fusion under model $*$
f_s	Sampling frequency of sensor (Hz)
$\mathcal{I}$	Cost of GP inference under model $*$
I_{sensor}	Current flowing through the sensor
k	Power/data rate coefficient
k_1, k_2	Positive battery decay coefficients
k_{env}	Flight adjustment coefficient
k_r	Propeller constant relating T with ω
k_{terr}	Terrain variation coefficient
m_R	Mass of robot
N_R	Number of rotors
O	Set of observed (input) locations
O_m	Set of observed (input) locations for the agent m
O_{global}	$\triangleq \{O_1 \cup O_2 \cup \cdots \cup O_M\}$ Super-set of observed (input) locations
P_{anc}	Power used by ancillary branch (W)
P_C	Power needed for computation & wireless communication (W)
P_{CP}	Net energy consumed by all components on-board
P_{FT}	Field trial power
P_{hover}	Power consumed while hovering
P_{IF}	Internal friction power
P_{man}	Power used by maneuvering branch (W)
P_{sensor}	Power consumed by the sensor
P_{terr}	Power consumed to overcome terrain resistance
PS_i	Pit stop i
R_*	Number of observations accounted under $*$
R_{motor}	Resistance of the motor
r_p	Radius of propellers (m)
R_{sensor}	Resistance of the sensor
s_0, s_1	Positive sensing coefficients (W)
T_{hover}	Thrust required to hover (N)
T_{test}	Execution time for the minimal load test

U	Set of unobserved (input) locations
U_m	Set of unobserved (input) locations for the agent m
U_{global}	$\triangleq \{U_1 \cap U_2 \cap \cdots \cap U_M\}$ Super-set of unobserved (input) locations
v	Forward velocity of robot (m/s)
v_{opt}	Optimal velocity which allows the robot to attain d_{max}
$\mathbf{M}$	Size of robot team
$\mathrm{tr}(\cdot)$	Trace of a matrix
$|\#(\cdot)|$	Cardinality of a matrix

Part I

The curtain raiser
A taster of this book

Contents

As the first part of this six-part book, this part provides a taster of what is included in this book. A brief overview of the chapters comprising this part is explained below.

I.1   Introduction

This chapter paints the big picture highlighting the theme of this book. It covers a wide range of applications, ranging from forest-fire to tsunami. It opens up potential research directions which could benefit from utilizing robots as such calamities take a toll on life of humans and live-stock.

I.2   Target environment

When utilizing robots for environment monitoring, it is essential to clarify the context in which the measurements can be taken. Broadly, the environment can be classified as aerial, ground, or marine, and this chapter mentions potential aspects of such environments that can be monitored within the limits of the current hardware.

I.3   Utilizing robots

This chapter mentions the classes of robots that can be utilized in the context of monitoring the target environments. Additionally, it talks about the categorization of the

available off-the-shelf sensors that can be mounted on the robots to monitor the target
environment.

1.4   Simultaneous Localization and Mapping (SLAM)

This chapter gives an overview of a conventional robotics "problem" called *Simultaneous Localization and Mapping (SLAM)*. In doing so, it introduces the concepts of
mapping, localization, and SLAM to the readers as some of these will be used later.

Introduction
What to expect?

1

If you are one of those people who like to get a taste of what they are about to sign-up for, read on.

Dr. Kshitij Tiwari

Contents

Highlights

- Statistics of global emission rates across nations by UN Environment Programme (UNEP)
- Natural disasters caused by ignorance towards environment
- Alarming need to monitor and manage environment for sustainability
- Conventional methods for environment monitoring
- Utilizing robots for efficient/intelligent monitoring

The global emissions are taking a significant toll on the environment and its dwellers, as is evident from some of the recent natural disasters. This chapter highlights some of the aftermaths of environmental imbalance caused by global emissions and discusses the motivation for monitoring the environment for ensuring sustainability.

1.1 Recapitulation of global emissions

According to the Paris Agreement [1], nations need to boost their efforts to keep the global warming below the $2\,°C$ mark [2]. The global emissions have reached a historic rate of $53.5\ GtCO_2e$ and continue to rise, which goes without saying that the national commitments to combat climate changes come up short. What once used to be lush green vegetation is now covered by disposed-off materials in most places, creating a landfill site as shown in Fig. 1.1. It is sites like these that further add to growing environmental concerns.

Figure 1.1 Negligence towards proper waste recycling transforms a lush green patch of land into a garbage landfill. Photo by Stijn Dijkstra from Pexels [3].

1.2 Emissions take a toll on nature

The United Nations Environment Programme (UNEP) defined "emission gap" as the comparison between the current levels of rise in temperatures to that during the pre-industrial periods. The current objective is to keep this emission gap to below the 2 °C mark [2] which is aimed to be achieved by 2030, but given the current commitments of the nations towards environmental emissions, it is looking highly unlikely that these goals can be met at the current rate.

To put things into perspective, this emission gap does not simply lead to melting of snow caps/glaciers. The recent California wildfires (Fig. 1.2) and tsunami in Indonesia (Fig. 1.3) are all caused to some extent by the imbalance in the nature stemming from the global emissions.

1.2.1 California wildfires

According to an article published in Aljazeera [4], more than 6000 wildfires have hit the USA as of 2018, which amounted to damages worth US$2.6bn. Some fires last well over a month and damage several thousands of hectares of arable land. As shown in Fig. 1.2, the smoke generated from such fires suspended over a significant area, and engulfed several territories along the west coast. The Mendocino Complex fire, the largest in the state history, had been tamed only after 35 days while two other such hotspots remain active in Northern California. As stated in the article [4] by Professor Noah Diffenbaugh, these trends are consistent with the global emissions, which in turn cause rise in temperature, leading to dryness and add fuel to such forest fires.

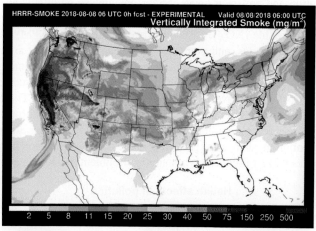

Figure 1.2 Illustrating the spread of smoke from the recent wildfires in California as published by CBS News. The blaze engulfed 283, 000 acres and still continues to grow. Image published by National Weather Service Office, San Diego [5].

1.2.2   Tsunami

On December 22, 2018, a tsunami ensued following an eruption of the Anak Krakatau volcano in the Sunda Strait and struck several coastal regions of Banten in Java and Lampung in Sumatra, Indonesia. The aftermath in the coastal town of Banten is apparent from Fig. 1.3 which shows the destruction at the Carita beach.

Figure 1.3 The aftermath in Carita Beach, Banten after a tsunami struck in 2018 [6].

According to [7], the death toll crossed 400, with 40,000+ people getting displaced and 14,000+ injuries.

1.3 Pollution takes a toll on human health

Not only Mother Nature, even its dwellers feel the brunt of imbalance caused by increasing pollution and emission levels. Major greenhouse pollutants also known as aerosol pollutants include carbon dioxide (CO_2), carbon monoxide (CO), sulphur oxide (SO_X), nitrous oxide (NO_X), suspended particulate matter (SPM), lead aerosol, volatile organic compounds, and other toxic materials. These gases are mainly produced by industries, automobile, agriculture activities, and even ordinary homes [8]. From different studies, it is well documented that when human beings come in contact, these chemicals/pollutants have adverse effects on human health. These chemicals are responsible for diseases, like lung cancer, pneumonia, asthma, chronic bronchitis, coronary artery disease, and chronic pulmonary disease. Fig. 1.4 represents the health effects due to pollution, such as air pollution, water pollution, and soil contamination, and demonstrates that air pollution is a major environmental risk to health.

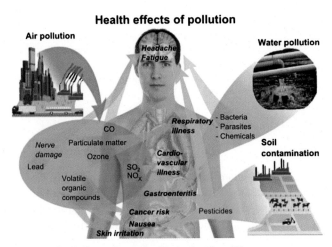

Figure 1.4 The impact of pollution on human health as taken from [8].

1.4 Need for environment monitoring

It is disheartening to see the loss of life, property, and livestock that ensues the natural disasters. It is equally alarming to see the suffering that ensues aerial, marine, and land contamination. If the causal nature of global emissions on instigating such natural disasters is well understood and paid heed to, such disasters and health risks can either be fully averted or at least, an early warning system can be developed. Besides, the environment has incurred severe damages, and being able to maintain and monitor what is left of it can go a long way in ensuring the sustainability of the species. The real challenge herewith is twofolds:

(i) To acknowledge the fact that while events like forest fires and tsunamis are visible and hence can be tackled, the underlying dynamics of the environment are largely unknown. On top of this, their interactions with inhabiting species further pose a research challenge which can be summed up as follows:

How are the environmental variations modeled and how do they vary across space and time?

(ii) The expanse of the areas (aerial, land, and marine) that need to be monitored to ensure sustainable living is significantly large, making it practically challenging to cover every nook-and-cranny effectively.

While the major part of this book addresses the former, the latter can be aided by the utilization of robots which have seen significant development in the recent decades. One such application of robots on a small scale is shown in Fig. 1.5, which illustrates an autonomous surface vehicle (ASV) that can be used to monitor marine life, harmful algal bloom, oil spill or similar applications. Similar robot platforms can also be utilized to monitor landfill scenarios as was previously illustrated in Fig. 1.1.

Figure 1.5 Solar-powered autonomous surface vehicle for marine environment monitoring. Image courtest of Liquid Robotics.

1.5 Conventional methods

In the past, some attempts have been made at monitoring the environments by deploying/positioning static sensors. This setup potentially looks like in Fig. 1.6. There are several challenges to such a setting:

- The sensing range of the sensors is not well defined, i.e., it is not always valid under real-world conditions to have a well-defined sensing range like a circular region with known radius.
- The number of sensors required to achieve a high spatial resolution, i.e., maximal area coverage for effective monitoring is infeasibly high.
- Even if such a large sensor array could be arranged, the vegetation cover and residential infrastructure obstructs the placement of such sensors at all desired locations.
- It is quite challenging to decide where to position the sensors.
- Owing to their "static" nature, the sensors cannot be moved as often as one would like to, to ensure optimal coverage at all times. This becomes quite restrictive.

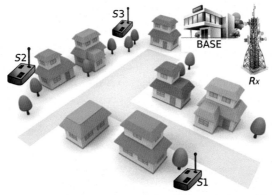

Figure 1.6 Illustrating static sensor placement for monitoring air quality in a residential area. The sensors are shown with $S1, \ldots, S3$, which are wireless transmitters picking up air quality measurements from the residential area. These measurements are then received by a receiver (Rx) which then transmits the data via wired connections to a processing site or base.

Owing to the aforementioned challenges in utilizing static sensor arrays for monitoring environments, this book addresses the use of robots as mobile sensor nodes which have higher agility and better area coverage capabilities.

1.6 Book organization

The aim of this book is to propose solutions for efficient deployment of robots for intelligent environmental monitoring which is done in six parts, each comprising of several chapters. The overview of each of these parts is as follows:

1. *Part-I (The curtain raiser)* is the introductory part, which discusses how this book came into being (Chap. 1), target environments that can be monitored (Chap. 2), brief discussion and categorization of robots (Chap. 3), and, a recapitulation of Simultaneous Localization and Mapping (SLAM) in Chap. 4.
2. *Part-II (The essentials)* is designed as a primer to equip the readers with the necessary jargon and mathematical models that will be utilized heavily in the latter parts

of the book. This part encompasses Chap. 5 which is an exhaustive overview of Bayesian Inference, multi-variate Gaussian distributions, gradient descent methods, and some mathematical gimmicks and optimization techniques utilized for inference that will follow suit. Chap. 6 then presents an extensive discussion about Gaussian processes (GPs) which are the non-parametric Bayesian regression method of choice for environmental monitoring widely used today. This chapter builds on the very basics of multi-variate Gaussian distributions and extends them to GPs and is targeted at novice and expert readers alike. Chap. 7 discusses several path planning algorithms that are primarily concerned with maximal area coverage; this is contrasted in Chap. 8, presenting another group of algorithms that are concerned with maximal information acquisition. By this time, the readers have been exposed to all the necessary technical and non-technical tools that would be required to ensure smooth transition across the rest of this book, some parts of which can get quite involved.

3. *Part-III (Mission characterization)* connects previously presented different independent components to form a coherent problem definition. This is done in Chap. 9 which explains the problem formulation. Aside from this, potential mission termination conditions are discussed primarily in terms of endurance/energy estimation setting (Chap. 10) and range estimation setting (Chap. 11).

4. *Part-IV (Scaling to multiple robots)*. Up until this part, most of the algorithms developed and discussed were potentially scalable to multiple robots, but this part is where multi-robot teams are explicitly discussed for the first time. Chap. 12 discusses the challenges and design considerations when scaling the team to multiple robots, and Chap. 13 then describes the challenge of maintaining a globally consistent model if multiple models are developed during run-time.

5. *Part-V (Continuous spatio-temporal dynamics)* primarily focuses on incorporating the temporal evolutions of the spatial dynamics and the design considerations that are impacted owing to this choice. These details are explained in Chap. 14.

6. *Part-VI (Epilogue)* is the last of this six-part book, which encompasses some real-world success stories as highlighted in Chap. 15, which describes robot fleet deployment for marine environment monitoring. Similarly, some significant success has been achieved in cumulus cloud monitoring, and one such project is described in Chap. 16. As an alternate, yet correlated application, Chap. 17 discusses the advances in search-and-rescue utilizing robots and some of author(s)' suggestions for the future research directions where GPs can also be incorporated herewith beyond just environment monitoring as a novel application consideration. A key challenge when using GPs and multi-robot teams to monitor large-scale environments is to decipher the robot location based on accrued measurements (signal-strength based localization) and this is described in Chap. 18. As a conclusion, the last chapter (Chap. 19) recapitulates the contents presented in this book, especially highlighting the author(s)' contributions and discusses potential future research directions.

For the convenience of the readers, each chapter has its respective bibliography at the end. All figures, if reproduced, have been aptly cited with due permissions from respective copyright owners. A cumulative list of reproduced material can be found at the end of the book.

References

[1] J. Rogelj, M. Den Elzen, N. Höhne, T. Fransen, H. Fekete, H. Winkler, R. Schaeffer, F. Sha, K. Riahi, M. Meinshausen, Paris Agreement climate proposals need a boost to keep warming well below 2 C, Nature 534 (7609) (2016) 631.

[2] United Nations Environment, Emissions Gap Report 2018, 2018.

[3] https://www.pexels.com/photo/photo-of-plastics-near-trees-2583836/.

[4] K. Corriveau, California wildfires damage in 2018 worth over $2.6bn, Aljazeera. URL https://tinyurl.com/y7sllm9v.

[5] National Weather Service, San Diego, Service, N.W. Wildfire smoke from California has reached New York City, 3,000 miles away. Twitter. URL https://tinyurl.com/yb8ct96x.

[6] https://en.wikipedia.org/wiki/2018_Sunda_Strait_tsunami#/media/File:Sunda_strait_tsunami_2.jpg.

[7] https://www.thestar.com.my/news/regional/2018/12/31/number-of-injured-in-indonesia-tsunami-surges-to-over-14000/.

[8] A. Kumar, H. Kim, G.P. Hancke, Environmental monitoring systems: a review, IEEE Sensors Journal 13 (4) (2013) 1329–1339.

Target environment
What can be monitored?

2

Earth provides enough to satisfy every man's need but not every man's greed.

Mahatma Gandhi

Contents

Highlights

- Illustrations of aerial, ground, and marine environments that can be monitored
- Discussion about types of measurements to be acquired
- Types of predictions defined for the scope of this book

The word "environment" is very generic and takes on a lot of forms in respective domains of research. Through this chapter, the notion taken by this word is described within the scope of this book, and some potential applications are discussed. The environment is primarily classified as per the nature of pollutants: freely suspended aerosols pollute the aerial environment, algae pollute the marine environment, while chemical spills pollute the ground environments. Additionally, the type of observations gathered are discussed, i.e., it could be useful in some cases to gather more information from areas with high uncertainty in pollution levels (could also stem from the uncertain nature of the pollutant or a combination thereof). While, in other cases, it could be instead beneficial to monitor areas with higher levels of pollution. Aside from these, terminologies are coined to classify the various types of predictions that can be made based on the model being used and the nature of the phenomenon being monitored.

Multi-Robot Exploration for Environmental Monitoring. https://doi.org/10.1016/B978-0-12-817607-8.00014-9

2.1   Aerial environments

Monitoring of contaminants suspended or dispersed in the air is critical since humans breathe such polluted air directly. Owing to the nature of aerosol pollution which is usually suspended in the air, Unmanned Aerial Vehicles (UAVs) serve as the best monitoring mechanism. They can be simply retrofitted with lightweight sensors and can cover large areas with considerable ease. A typical aerial robot that could be useful for this purpose is shown in Fig. 2.1. In [1], the authors have tried to model the boundaries of the contaminant cloud and predict its dispersal to quickly contain it. Similar efforts have been carried out in works like [2], wherein the authors described gas-detection sensors that are mountable on small UAVs. In [3], the researchers investigated if it is possible to utilize bioinspired models to account for airflow, which got rid of the assumption that the airflow is either uniform or unidirectional.

It is not just about monitoring pollutants that contaminate breathable air. Other factors to be monitored could include cloud formation. In [4], the researchers presented novel mechanisms to monitor the formation and evolution of low-altitude continental cumulus clouds. Such methods could help to reform the ways in which crop cultivation is planned at the moment. Monitoring aerial environments whilst obtaining uncertainty estimates can go a long way in predicting the optimal cropping patterns by taking into account the forecasted weather during the cultivation season [5]. An extensive survey of utilization of aerial robots for remote sensing has been presented in [6] which summarizes the typical platforms and on-board sensors used during real-life deployments.

Figure 2.1 DR1000 flying laboratory drone equipped with sensors for monitoring air quality. Image used with permission from Scentroid [7].

2.2   Marine environments

Monitoring marine pollution is crucial for protection and sustainability of the marine biology. For instance, after the oil spill in the Gulf of Mexico (shown in Fig. 2.2), 8332 different marine biology species got caught in its wake and were adversely affected. As of 2016, a research team reported that 88% of about 360 baby or still-born dolphins within the spill area "had abnormal or under-developed lungs", compared to 15% in other areas [8]. Similarly, excessive blooming of algae can produce extremely dangerous toxins that can sicken, or, in the worst case, kill people and marine life. This also increases the cost incurred for processing water to make it potable. Thus, researches like [9] have investigated the means of monitoring harmful algal blooming.

Aside from marine pollutants, the health of a marine ecosystem can also be monitored using marine robots. This is illustrated in works like [10] that utilized a heterogeneous team of mobile robots for assisting scientists in monitoring a marine ecosystem. Other works in the robotics domains are more focused on developing amphibious robots that can gather high resolution images of marine life as shown in [11].

Figure 2.2 Oil Spill in the Gulf of Mexico captured by the Moderate Resolution Imaging Spectroradiometer (MODIS), NASA. The oil spill is just off the coast of Louisiana and is highlighted in white. Image originally created by Jesse Allen, using data provided courtesy of the University of Wisconsin's Space Science and Engineering Center MODIS Direct Broadcast system. Used with permission from NASA Earth Observatory.

2.3   Ground environments

Ground environment monitoring encompasses deweeding applications like those shown in Fig. 2.3. Usually deweeding is a complicated challenge given the size of the farms to be monitored. Usually these problems are tackled by spraying chemicals that eradicate the weeds but the chemicals that remain on the edible plants are becoming a health concern. To this end, in [12] an organic deweeding mechanism was presented, aided by robots. As opposed to conventional organic methods of grasping and plucking the weeds, in this work the robot ingeniously smashed the weeds down into the ground.

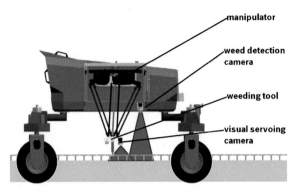

Figure 2.3 Robotic weed detection using a robot designed by Deepfield Robotics [12].

Aside from weed, the ground pollution also manifests itself in terms of soil contamination. This usually occurs with the soil naturally trapping potentially toxic trace elements (PTTEs) like heavy metals, metalloids etc. While entrapping such toxic elements prevents the pollution of the underlying water table, when the soil reaches critical loading levels, it is said to be contaminated/polluted [13].

2.4 Types of observations

In order to be able to monitor the variety of environments mentioned above, a robot needs to be told what it should observe and where in the target environment it can get those observations. The most common way to do this is to follow the information gradient. Intuitively, this means visiting the locations which have the highest amount of information. What then remains is to define a representation of information. As an illustrative example, consider Fig. 2.4 and assume that the map is the information map where the variety of peaks can be broadly classified as follows:

- *High measurements.* These locations represented by A in Fig. 2.4 exhibit very high values. This could, for instance, be the case in aerial pollution monitoring scenarios wherein the areas close to a chemical plant will have higher measurements of pollutants as opposed to areas in the suburbs.
- *High variance.* These locations represented by B in Fig. 2.4 exhibit high variance. As variance is a measure of uncertainty, these areas represent high uncertainty in pollution levels owing to unknown nature of pollutants, a combination thereof or similar reasons. Capturing measurements exhibiting high variance could be useful

in the monitoring of the fisheries where some fish species are very vulnerable and sensitive to even the slightest of changes in the environment.

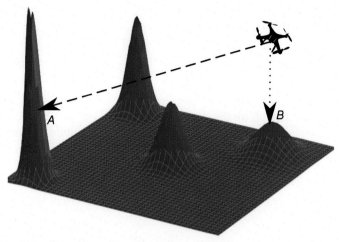

Figure 2.4 Acquiring observations with a single robot.

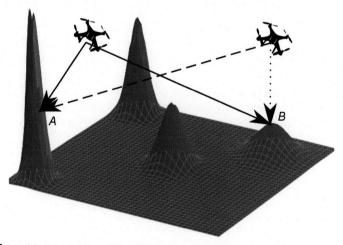

Figure 2.5 Acquiring observations with multiple robots.

The last challenge remaining is to select from the two possible types of observations and to make it clear to the robot which one it should be aiming for, while exploring the environment of interest. This falls under the domain of informative path planning about which further details follow suit in a later part (Chap. 8) of this book. Fig. 2.5 represents a similar setup but scaled up to multiple robots (only two shown for illustration). This is usually beneficial as the areas to be monitored are generally vast and cannot be covered by a single robot. This further complicates the problem of

How is the team coordinated? and *who goes where?* These challenges are addressed at length in Chaps. 8 and 12, respectively.

2.5 Types of predictions

In this section, primarily three kinds of predictions are summarized, depending on the nature of the query. They are:

- *Interpolation*, which refers to making predictions over a previously unobserved location or a set thereof, for the current time-step.
- *Extrapolation*, which refers to making predictions over future time-step(s) for a previously observed location or a set thereof.
- *Forecasting*, which refers to making predictions for previously unobserved locations for a future time-step.

Throughout this book, the term "prediction" or related terms primarily refer to *interpolation*.

2.6 Summary

This chapter discussed the primary classes of target environments that a robot could be presented with. For this, aerial, marine, and ground environments were described and related works were summarized. Along these lines, two classes of observations were also mentioned, *high measurements* and *high variance*, depending upon the application that can be used to decide the path-planning mechanisms for the robot. Also, three types of prediction scheme were defined of which *interpolation* will be the primary focus in this book. In the next chapter, several categories of robots and sensors will be discussed.

References

[1] B.A. White, A. Tsourdos, I. Ashokaraj, S. Subchan, R. Zbikowski, Contaminant cloud boundary monitoring using network of UAV sensors, IEEE Sensors Journal 8 (10) (2008) 1681–1692.

[2] J.A. Malaver Rojas, N. Motta, L.F. Gonzalez, P. Corke, A. Depari, Towards the development of a gas sensor system for monitoring pollutant gases in the low troposphere using small unmanned aerial vehicles, 2012.

[3] V. Hernandez Bennetts, A.J. Lilienthal, P. Neumann, M. Trincavelli, Mobile robots for localizing gas emission sources on landfill sites: is bio-inspiration the way to go?, Frontiers in Neuroengineering 4 (2012) 20.

[4] C. Reymann, A. Renzaglia, F. Lamraoui, M. Bronz, S. Lacroix, Adaptive sampling of cumulus clouds with UAVs, Autonomous Robots 42 (2) (2018) 491–512.

[5] D. Rodriguez, P. De Voil, D. Hudson, J. Brown, P. Hayman, H. Marrou, H. Meinke, Predicting optimum crop designs using crop models and seasonal climate forecasts, Scientific Reports 8 (1) (2018) 2231.

[6] G. Pajares, Overview and current status of remote sensing applications based on unmanned aerial vehicles (UAVs), Photogrammetric Engineering and Remote Sensing 81 (4) (2015) 281–329.

[7] http://scentroid.com/scentroid-dr1000/.

[8] J. Staletovich, Hundreds of baby dolphin deaths tied to BPGs Gulf oil spill, The Miami Herald (2016), URL https://www.myrtlebeachonline.com/news/nation-world/national/article71524207.html.

[9] D.M. Anderson, Approaches to monitoring, control and management of harmful algal blooms (HABs), Ocean & Coastal Management 52 (7) (2009) 342–347.

[10] F. Shkurti, A. Xu, M. Meghjani, J.C.G. Higuera, Y. Girdhar, P. Giguere, B.B. Dey, J. Li, A. Kalmbach, C. Prahacs, et al., Multi-domain monitoring of marine environments using a heterogeneous robot team, in: Intelligent Robots and Systems (IROS), 2012 IEEE/RSJ International Conference on, IEEE, 2012, pp. 1747–1753.

[11] G. Dudek, M. Jenkin, C. Prahacs, A. Hogue, J. Sattar, P. Giguere, A. German, H. Liu, S. Saunderson, A. Ripsman, et al., A visually guided swimming robot, in: Intelligent Robots and Systems, 2005 (IROS 2005). 2005 IEEE/RSJ International Conference on, IEEE, 2005, pp. 3604–3609.

[12] A. Michaels, S. Haug, A. Albert, Vision-based high-speed manipulation for robotic ultra-precise weed control, in: Intelligent Robots and Systems (IROS), 2015 IEEE/RSJ International Conference on, IEEE, 2015, pp. 5498–5505.

[13] E. Galán, A.J. Romero-Baena, P. Aparicio, I. González, A methodological approach for the evaluation of soil pollution by potentially toxic trace elements, Journal of Geochemical Exploration 203 (2019) 96–107.

Utilizing robots
Robots to assist with monitoring

Not all robots are terminators. They can also be used for the greater good.

Dr. Kshitij Tiwari

Contents

Highlights

- Categorization of various robots applicable to environment monitoring application
- Discussion about types of sensors to be mounted on robots
- Real-life example to illustrate impact of various sensing ranges

Humans have a mixed perception when it comes to robots and their utilization for a variety of tasks. While a vast majority of people see robots as a threat to their jobs, there is also a segment of the world population that believes the robots can help humans for the greater good. Lately, robots have been deployed in various domains like factory automation, medical diagnostics, surgeries, defense, search-and-rescue, and traffic management, just to name a few. Robots come in all shapes and sizes. Some are meant to be "cute" to make them suitable for human–robot interaction, whilst others, not so much. They might even look intimidating to some, especially if they are weaponized for defense and security applications. Depending on the design and operational considerations, the available mobile robots can be broadly classified into the following three categories: aerial, marine, and ground vehicles. While the details follow suit, it is essential to point out that there are other robot categories, like humanoids or industrial articulated manipulators, which do not directly contribute to the applications of interest of this book, and hence have been disregarded here. Most of these robots originally came into being to assist with access to dirty-dangerous-and-dull (DDD) [1] environments.

Aside from mechanical and aerodynamic considerations, sensors also play a key role. The better informed the robot is (via its sensors), the more informed decisions can be made. To this end, without going into explicit sensory details, sensors are broadly classified based on their sensing ranges. A real-life example is then discussed to illustrate how long- and short-range sensors affect decision making.

3.1 Unmanned Aerial Vehicles (UAVs)

Unmanned Aerial Vehicles (UAVs), or more commonly known as drones, are typically the robots that can fly (i.e., operate in aerial environments) and are meant to be operated without a human on-board: either autonomously using on-board computing resources, or via teleoperation by ground control staff. Unmanned vehicles were primarily designed for the military applications but, of late, their low-cost commercial counterparts are also available for civilian use. Depending on the structure of the wing and the principles of flight dynamics, drones can be further categorized as rotor-wing or fixed-wing (cf. [2]) aerial vehicles as shown in Fig. 3.1.

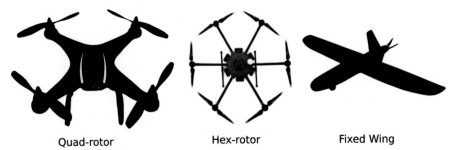

Quad-rotor   Hex-rotor   Fixed Wing

Figure 3.1 Various types of UAVs depending on their flight mechanism. Quad-rotor (left) has 4 rotor blades while a hex-rotor (center) has 6 rotor blades. As opposed to these mechanisms, there are fixed-winged gliders (right) which have fixed wings and have a contrasting aerial flight mechanism.

3.1.1 Rotary-winged UAVs

Rotary-wing UAVs consist of several rotor blades that revolve around a fixed mast, known as a rotor. Rotary wing UAVs also come in a wide range of setups consisting of a minimum of one rotor (helicopter), 3 rotors (tricopter), 4 rotors (quadcopter),

6 rotors (hexacopter), or 8 rotors (octocopter). Of late, manufacturers have scaled up the models to incorporate up to 16 rotors, but they are not prominently used, yet.

Rotor blades work by utilizing the rotors to produce airflow over the blades by constant movement which produces the required airflow over their airfoil to generate lift. Control of rotary UAVs comes from the variation in thrust and torque from the rotors, e.g., the downward pitch of a quadcopter is generated from the rear rotors producing more thrust than the forward rotors. This enables the rear of the quadcopter to rise higher than the front, thus producing a nose-down attitude. Yaw movement is controllable using the torque force of the rotors where the diagonal rotors either spool more or less than their counter-diagonal rotors, thus producing an imbalance in the yaw axis and causing the quadcopter to rotate on the vertical axis. Thus, by controlling the rotors aptly, a higher degree of agility can be obtained. This is illustrated in Fig. 3.2 wherein mg represents the weight of the drone, $F_{i \in \{1,...,4\}}$ represents the forces for the corresponding rotor i; ω_i represents the angular velocity of the corresponding rotor. The point labeled with gray sphere represents the center of mass and the origin of the rotor frame. The same can be scaled up or down to other rotor configurations.

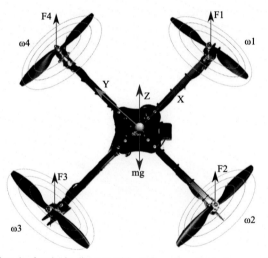

Figure 3.2 Illustrating the free-body diagram with all the active forces for a quadrotor. Dynamics are controlled by managing the interaction of these forces. Image taken from [3].

Such agility and control comes at a cost, which is not all negative but has both an upside and downside. Below, the pros (+) and cons (−) of rotor-winged UAVs are summarized:

+   Can take-off and land vertically doing away with the need for runways
+   Can operate even in confined spaces owing to their small size and agility provided by rotor-wings
+   Can hover and perform agile maneuvers owing to several degrees of freedom provided by rotor-wings
−   Design and flight control are often more complex than for simple fixed-winged aerial vehicles

— Have significantly shorter flight durations (of the order of minutes) which constraints the size of the coverage area

It can be argued that flight time is being improved by developing new designs that rely on natural resources like wind and solar power to increase the endurance as opposed to conventional power sources. However, such solutions are not yet commercialized/standardized for widespread usage.

3.1.2 Fixed-winged UAVs

Fixed-wing UAVs consist of a rigid wing that has a predetermined airfoil as shown in Fig. 3.3, which generates lift caused by the UAV's forward airspeed. Although the figure illustrates the concept for an airplane, but the same four forces are in play even for smaller fixed-wing UAVs. This airspeed is generated by forward thrust usually by the means of a propeller being turned by an internal combustion engine or electric motors. Control of the UAV comes from control surfaces built into the wing itself, which traditionally consist of ailerons, an elevator, and a rudder. They allow the UAV to freely rotate around three axes that are perpendicular to each other and intersect at the UAV's center of gravity. Specifically, the elevator controls the pitch (lateral axis), ailerons control the roll (longitudinal axis), and the rudder controls the yaw (vertical axis).

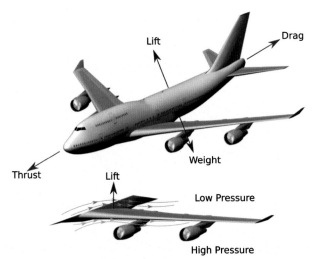

Figure 3.3 Illustrating the free-body diagram with all the active forces for a fixed-wing UAV. Owing to the difference in air pressure above and below the wing, a lift is created which allows the UAV to stay afloat. Drag represents the aerial resistance and thrust represents the force that propels a flying machine in the direction of motion while overcoming the aerial drag.

Yet again the design sounds simple enough, but this comes at a cost as well. The pros and cons of fixed-wing UAVs are:

+   Simple structure

+ Easy to maintain and repair
+ Efficient aerodynamics providing for prolonged flight durations at high speeds
+ Longer flight times allow for larger surveillance area coverage
+ Can perform natural gliding like an eagle floating in the air under strong wind gusts without any power
+ Allow for carriage of heavier payloads
− Need a runway to attain take-off speeds or a launcher as is the case of take-off from naval carriers
− Need for runway can be aided by the use of vertical take off/landing (VTOL) and short take off/landing (STOL) solutions which are not readily available in smaller sizes for small-scale civilian applications
− Need to constantly keep moving forward to stay afloat which means that they cannot hover, making them unsuitable for tasks like inspection

Given that UAVs operate in aerial environments, significantly higher than the ground planes, they have the advantage of providing a wider field-of-view (fov). This could be useful in applications like wildfire and flood monitoring scenarios where they can provide a better picture for post-disaster scenarios.

3.2 Unmanned Marine Vehicles (UMVs)

Unmanned Marine Vehicles (UMVs) have a wide variety of applications in marine geosciences: military, scientific, environmental and commercial. They can be broadly classified as surface vehicles referred to as Autonomous Surface Vehicles (ASVs) similar to Fig. 1.5 or underwater vehicles referred to Autonomous Underwater Vehicles (AUVs) as shown in Fig. 3.4. They are often used to gather knowledge about the extreme environments like hydrothermal vents or marine life under the thick layers of polar ice. These regions are inaccessible to humans and thus the ability for such robots to withstand harsh conditions comes in handy. These robots are primarily meant to operate in marine environments and provide widespread data like seabed mapping, coral-reef blooming, marine life health, and related information.

3.2.1 Underwater vehicles

When it comes to monitoring the previously impenetrable environments such as the sea surface, the deep sea and under sea ice, unmanned underwater vehicles (UUV) alternatively referred to as AUVs are the platform of choice. Although there are over 75 such platforms [5], there are several pros and cons in general pertaining to this platform. They are:

+ small and quiet
+ low cost
+ robust to changes in weather conditions
+ help in gathering data to monitor deep sea ecosystem without putting scientists and marine biologists at risk

Figure 3.4 Illustrating the UK Natural Environment Research Council (NERC) Autosub6000 AUV that is useful for marine research [4].

+ capable of monitoring marine ecosystem without affecting their natural behavior
− have limited deployment range
− care must be taken while resurfacing to avoid collision
− deployment and recovery must be done via a mother vessel which is challenging during harsh weather
− needs to be contact with base station/remote operator but water hinders/distorts the transmission

3.2.2 Surface vehicles

Sometimes marine biologists and oceanographers require bathymetry measurements which are acquired in shallow, near-shore environments to estimate the potential distribution of harmful algal bloom which contribute to marine pollution. Similarly, ocean current observations near icebergs and marine-terminating glaciers are necessary to be monitored to estimate their melt rates. Doing so, using manned vehicles poses a risk so unmanned surface vehicles are being increasingly deployed [6]. Some of the challenges faced with the underwater vehicles can be addressed by using surface vehicles which have their own strengths and weaknesses as discussed.

+ usually lighter than underwater vehicles
+ owing to surface operations, allow for satellite links providing larger communication range
+ easily deployable despite harsh weather conditions
+ owing to surface operations, allow for utilization of alternative power sources like solar power etc. to proper the vehicle
− owing to surface operations, they must be always prepared to interact and avoid other shipping vessels

— owing to their nature, they are rendered useless for sub-surface operations and hence, must always tackle waves to stay afloat

Thus, depending on the nature of deployment, sometimes it is benefitial to use AUVs whilst other times ASVs come to aid.

3.3 Unmanned Ground Vehicles (UGVs)

Unmanned Ground Vehicles (UGVs) have been used for ground based operations primarily by the military for bomb defusal or reconnaissance of hostile territories. Of late, owing to commercially available platforms, UGVs are also being used for civilian research and leisure as well. Based on the terra-mechanics models, these robots can be broadly classified into wheel- or track-based robots, and an example of each category is shown in Fig. 3.5.

Tracked Robot
(A)

Wheeled Robot
(B)

Figure 3.5 Illustrating the two main categories of ground robots: (A) robots using tracks for locomotion (image taken from [7]) and (B) robots using wheels for navigation (image taken from [8]).

The design of the locomotion systems of mobile robots for unstructured environments is generally complex, particularly when they are required to move on uneven or soft terrains, or to climb obstacles. Thus, it becomes crucial to choose between wheeled and tracked robots for the suitable application. The next sections present pros and cons of each of these designs which are mostly inspired from the review article presented in [9].

3.3.1 Wheels

The wheels are the most common structure for a robot aimed at simplicity, agility, swiftness, ease of controllability, and a lot more. This section provides a brief review of wheeled locomotion systems highlighting their pros and cons:

+ Easy repair/maintenance using 3D-printing
+ Low cost of production. They can even be 3D-printed
+ Low torque needed to break over the static friction barrier and set off the motion from stationary position

+ Higher maneuverability with ability to spin on the spot by controlling motor directions accordingly
+ Lightweight and simple design
+ Variety of materials can be used to build wheels depending on mission characteristics
− Ground clearance is challenging when maneuvering an obstacle course with tall obstacles
− High risk of wheels getting trapped within crevices of uneven surfaces
− Low traction on the ground

3.3.2 Continuous tracks

The continuous band of treads driven by a series of wheels is used when the wheels cannot be used (e.g., post-disaster scenarios which require maneuvering over uneven rubble as showcased in [10]). Thus, despite several pros of wheeled locomotion systems, they are largely rendered useless in such settings or when a robot may even be required to climb vertically, often requiring high traction which cannot be obtained by wheeled systems. This is where the tracked locomotion system comes to the rescue, but at some costs associated with it. Below, the pros and cons of tracked systems are presented:

+ Tracked systems are often more power-efficient compared to wheeled and even deliver higher performance
+ Higher traction over wheeled systems which allows for easy locomotion on hard-to-maneuver surfaces like ice, snow, rubble, vertical obstacles, walls, etc.
+ Lower pressure exerted on the surface of operation. This is extremely useful in cases where the surface itself is unstable like post an earthquake, building collapse, fire, etc.
+ Given the weight distribution across the entire track, the robots operating on tracked systems have a higher payload capability
− Continuous nature of the tracked system means that even the slightest of faults in any connecting joint would render the track unusable, bringing the robot to a staggering stop
− Reduced maneuverability and inability to turn in place or perform precise maneuvers
− Fragile nature and short life-spans
− Frequent and challenging repair and maintenance needs

3.4 Sensors

Just like a human needs sensory feedback to be able to perform tasks and interact with the environment, a robot too needs sensory capabilities, howsoever rudimentary they may be. While there are a variety of sensors available these days, for the scope of this work, the sensors are broadly classified into two categories based on the range they cover and the amount of information they acquire.

3.4.1 Sensor range

Each sensor has a specific region around, which it can cover and provide feedback for. This is what is referred to as the *sensing range*. In robotics, it is a common practice to model the sensing range as a Gaussian distribution of the form $R \sim \mathcal{N}(0, \sigma)$ where R represents the sensing range which is distributed as a zero-mean Gaussian with i.i.d. (or independent and identically distributed) variance. Depending on the spread of the variance of the sensing range, these sensors either fall under the long- or short-range sensing criteria.

3.4.1.1 Short-range sensors

Sensors like sonars, bumpers, whiskers, tactile sensors, etc., fall under the short-range sensing category given the limited sensing range that provides only local coverage.

3.4.1.2 Long-range sensors

As opposed to short-range sensors, sensors like camera, laser-range finder, LIDAR, sonar, etc., fall under the long-range sensing category given the wider fov.

3.4.2 Sensory information

Based on the amount of information accrued, the sensors can be further classified as follows.

3.4.2.1 Interoceptive sensors

The interoceptive sensors, also known as (a.k.a.) point-sensors, only provide sensory information about the variable of interest at the sensor location itself. In other words, the sensor coverage area is a point overlaying the sensor location itself. Such situations are most commonly alluded to in the environment monitoring literature as the robots/sensors can only perceive key variables at the positions they can reach.

3.4.2.2 Exteroceptive sensors

The exteroceptive sensors are the mostly commonly used sensors in the robotics domain. These sensors cover a wide area around the sensor and help accrue a significant amount of data samples at a time.

3.5 Real-life example

Consider a real-life example shown in Fig. 3.6 to get an intuitive understanding of the limits of short- and long-range sensing. This figure shows as autonomous driving car equipped with an LIDAR and sonar sensors. Although a real autonomous driving car may have additional sensors like an RGB-D camera, for the sake of simplicity, they have been disregarded in this discussion.

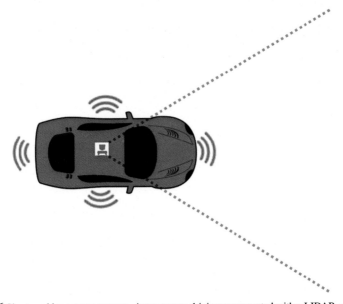

Figure 3.6 Short- and long-range sensors. Autonomous driving car mounted with a LIDAR and sonar.

While the sonar (short-range) sensor provides the information about the presence or absence of obstacles in the immediate vicinity of the car, this information in itself is very limited. For instance, this cannot be used to plan the behavior of the car several steps into the future or to plan for contingencies for unforeseen scenarios. As opposed to this, an LIDAR gives a much wider range of information spread over a larger region which can be used to have a multitude of information for processing.

3.6 Summary

This chapter introduced several classes of robots and discussed some of the potential operational environments that are best suited given their locomotion models. Aside from those discussed herewith, there are other categories of robots that do not fit the scope of this book and hence have been left out of discussion for brevity. As robots need sensors to be aware of the operational conditions, a brief overview of generic sensors was presented by categorizing them based on their coverage potentials. Towards the end, this chapter presented a real-life example of an autonomous driving car equipped with both short- and long-range sensors and analyzed how the various sensor classes affect the decision making strategies while operating. In the next chapter, a category of research problem referred to as *SLAM* will be introduced, which is most commonly faced by robots operating in unknown environments.

References

[1] R. Murphy, R.R. Murphy, R.C. Arkin, Introduction to AI Robotics, MIT Press, 2000.

[2] M. Boon, A. Drijfhout, S. Tesfamichael, Comparison of a fixed-wing and multi-rotor UAV for environmental mapping applications: a case study, in: The International Archives of Photogrammetry, Remote Sensing and Spatial Information Sciences, vol. 42, 2017, p. 47.

[3] https://commons.wikimedia.org/wiki/File:Quadrotor.jpg.

[4] R.B. Wynn, V.A. Huvenne, T.P. Le Bas, B.J. Murton, D.P. Connelly, B.J. Bett, H.A. Ruhl, K.J. Morris, J. Peakall, D.R. Parsons, et al., Autonomous underwater vehicles (AUVs): their past, present and future contributions to the advancement of marine geoscience, Marine Geology 352 (2014) 451–468.

[5] P.G. Fernandes, P. Stevenson, A.S. Brierley, AUVs as research vessels: the pros and cons, Ices Cm (2002).

[6] D.F. Carlson, A. Fürsterling, L. Vesterled, M. Skovby, S.S.Pedersen, C. Melvad, S. Rysgaard, An affordable and portable autonomous surface vehicle with obstacle avoidance for coastal ocean monitoring, Hardwarex 5 (2019) e00059.

[7] https://tinyurl.com/y32xgur7.

[8] https://commons.wikimedia.org/wiki/File:KUKA_youBot.jpg.

[9] L. Bruzzone, G. Quaglia, Locomotion systems for ground mobile robots in unstructured environments, Mechanical Sciences 3 (2) (2012) 49–62.

[10] W. Lee, S. Kang, M. Kim, M. Park, ROBHAZ-DT3: teleoperated mobile platform with passively adaptive double-track for hazardous environment applications, in: Intelligent Robots and Systems, 2004 (IROS 2004). Proceedings. 2004 IEEE/RSJ International Conference on, vol. 1, IEEE, 2004, pp. 33–38.

Simultaneous Localization and Mapping (SLAM)

Where is the robot?

You don't know what you don't know.

Socrates

Contents

Highlights

- Overview of Mapping
- Overview of Localization
- Conventional SLAM
- Bio-inspired SLAM

This chapter is primarily aimed at reviewing some of the notable works in the domain of Simultaneous Localization and Mapping (SLAM). SLAM is a situation where the robot starts off in an unknown environment, and is tasked with exploring it to learn the environmental structure (mapping) while simultaneously deciphering its own location (localization) with respect to the map being generated. This clearly is a chicken-and-egg problem wherein the performance of either component affects both peers. It is arguable whether SLAM is a research problem in itself or a systematic approach to solve a research problem of navigating in unknown environments [1,2].

Multi-Robot Exploration for Environmental Monitoring. https://doi.org/10.1016/B978-0-12-817607-8.00016-2

4.1 Mapping

The task of exploring the environment and simultaneously learning its spatial representation is called *mapping*. Depending on the nature of the information that gets encoded in the maps, they can be of several types and the most relevant ones are summarized below.

4.1.1 Metric maps

Metric maps are the most commonly used and human readable form of map, which record geometric configuration of spaces. It goes without saying that each object is mapped with its precise coordinates such that the map is view-point invariant [3], yet prone to noise. This is easy to generate and is very human-readable. An illustration is shown in Fig. 4.1.

Figure 4.1 A metric map generated using Octomap [4] for Freiburg campus.

4.1.2 Topological maps

Topological maps are a graph-based framework, which only encode relationships between objects (represented as nodes) as shown in Fig. 4.2. The edges between such nodes represent the relationships such as distances in real world. This is a compressed representation of the world, thereby it is efficient in terms of memory overhead.

4.1.3 Topometric maps

In [6], an integrated framework was presented that maintains a global topological map by connecting several local metric maps. This allows for compact map representation whilst getting rid of the need for having a global map consistency, making the

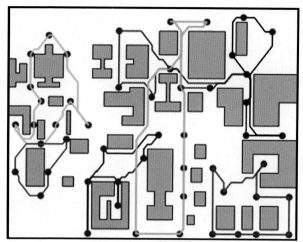

Figure 4.2 A topological map as presented in [5]. The picture shows the trajectories of 7 robots for the patrolling of a part of the UCSB campus. The viewpoints (red ●) are chosen to provide sensor coverage of the whole area.

approach robust and precise. An illustration of such a topometric map is shown in Fig. 4.3. In [7], the authors presented a vision-based hierarchical topometric map that builds a globally-consistent pose-graph with locally consistent point clouds attached to each node.

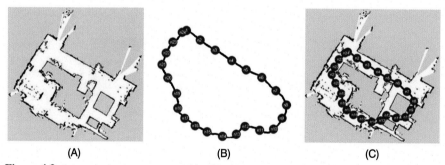

        (A)   **(B)**   **(C)**

Figure 4.3 A topometric map as presented in [8]: (A) metric map of the area explored and generated using laser scanner; (B) corresponding topological map of the explored area; (C) the hybrid topometric map, where each node in the topological graph is registered with a spatial specific region of the occupancy grid.

4.1.4 Semi-metric topological maps

In [1, Chap. 10], a novel semi-metric topological map called the Experience map was introduced. Compared to a topological map, the nodes here encode the position information (metric scale) and orientation information (non-metric representation in degrees), and hence, the name.

4.1.5 Measurement maps

In [9], the authors presented techniques for modeling ambient magnetic fields by utilizing non-parametric Bayesian inference methods called Gaussian Processes (GPs). The map as shown in Fig. 4.4 is referred to as a measurement map and represents the continuous measurement that can be queried at any spatial location of interest.

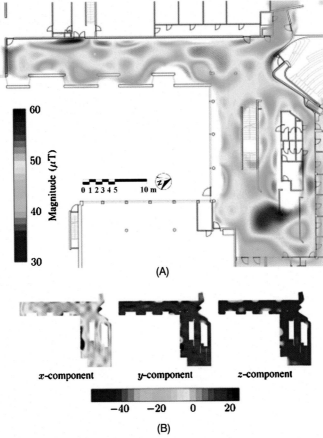

Figure 4.4 A measurement map as presented in [9]; (A) interpolated magnetic field of the lobby of a building at the Aalto University campus; (B) vector field components.

4.2 Localization

Localization is the problem of inferring the location with respect to a map. When dealing with a pure localization problem, it is often assumed that the metric map of the target environment is known *a priori*. The challenge then is to identify where in the map the robot is.

4.2.1 Global localization

Global localization, a.k.a. *kidnapped robot* or *pose estimation* problem refers to the task of localization (pose estimation with respect to the global frame of reference) with vague or no prior knowledge of the initial pose [10]. In the kidnapped robot problem, the robot is teleported to an arbitrary, yet perceptually aliased scene in the target environment. Owing to teleportation, the robot does not get odometric feedback, and the visual ambiguity leads the robot to believe that nothing has changed. However, as the robot starts to move around, its belief does not align with the visual feedback which is when the robot is "lost". Although the kidnapped robot problem sounds very artificial, given that in reality almost never does a robot gets teleported like this, it aims at testing the recovery capability of localization systems in case of extreme failures [11].

Some approaches to solving this problem include Monte Carlo localization [12] or received signal strength based localization [13], the details of which have been omitted here for brevity. In [14], Monte Carlo localization methods were presented utilizing metric maps. In [15], the authors presented Monte Carlo localization techniques relying on the measurement maps.

4.2.2 Local localization

Local localization is simply a pose-tracking problem wherein the robot is *a priori* made aware of its starting pose and is tasked with tracking its pose as it moves. This is quite different from the *pose estimation* problem [10] discussed above, given the prior knowledge of initial pose being made available here, which was not the case above. In [16], a pose tracking algorithm is presented which relies on parametrized models of the target (metric) environment and utilizes a type of Bayesian filter called Kalman filter [17,18] for recursive state estimation.

4.3 Simultaneous Localization and Mapping (SLAM)

Simultaneous Location and Mapping (SLAM) has been a core area of research interest for robotics researchers, as is evident from [19] and [20]. As of today, SLAM can be categorized into two categories, conventional SLAM which includes artificial solutions (probabilistic models) to addressing SLAM and bio-inspired SLAM which relates to solutions that are inspired by neural substrates observed in mammals and insects.

4.3.1 Conventional (probabilistic) SLAM

Consider a set of sensory observations $o^{[0:t]}$, up until the current (discrete) time-step, t. Then, the SLAM problem can be formulated as estimating the current state of the robot x^t and, simultaneously, the current estimate of the map of the environment, m^t. The state of the robot in this setting refers to the estimated pose at time t. Mathematically, this entails computing the joint probability, $p(m^t, x^t | o^{[0:t]})$. Using the Bayes rule, this results in iteratively calculating the location and map posteriors as follows:

$$\underbrace{p(x^t | o^{[0:t]}, m^t)}_{\text{location posterior}} = \frac{1}{\zeta_1} \sum_{m^{t-1}} \underbrace{p(o^t | x^t, m^t)}_{\text{likelihood}} \sum_{x^{t-1}} \underbrace{p(x^t | x^{t-1})}_{\text{transition}} \underbrace{p(x^{t-1} | m^t, o^{[0:t-1]})}_{\text{prior}},$$

$$\underbrace{p(m^t | x^t, o^{[0:t]})}_{\text{map posterior}} = \frac{1}{\zeta_2} \sum_{x^t} \sum_{m^t} \underbrace{p(m_t | x_t, m_{t-1}, o_t)}_{\text{likelihood}} \underbrace{p(m_{t-1}, x_t | o_{1:t-1}, m_{t-1})}_{\text{prior}}.$$

$$(4.1)$$

Eq. (4.1) describes the recursive Bayesian update rule for both the location and map posteriors with respect to their corresponding likelihood and prior probability distributions. Likelihood models are often also referred to as *observation models* while the transition model $p(x^t | x^{t-1})$ is also known as the *motion model*.

Putting it all together,

$$\underbrace{p(m^t, x^t | o^{[0:t]})}_{\text{slam posterior}} = \frac{1}{\zeta_3} \underbrace{p(x^t | o^{[0:t]}, m^t)}_{\text{location posterior}} \times \underbrace{p(m^t | x^t, o^{[0:t]})}_{\text{map posterior}} \qquad (4.2)$$

gives an iterative update rule for SLAM in terms of the localization and map posterior, each of which gets updated as the new observations are acquired. In the equations above, ζ_* represent the corresponding normalization constants.

4.3.2 Bio-inspired SLAM

Some researchers are also starting to explore the niche area of bio-inspired SLAM. Although not much is known about the human brain, the rat brain, fortunately, has been significantly studied. This gave rise to a bio-inspired SLAM algorithm called RatSLAM [1, Chap. 8] which looks at the neural substrates that assist with spatial navigation in rats. These mechanisms are then adapted to assist with robotic tasks like navigation, mapping and eventually SLAM.

4.4 Summary

This chapter introduced the chicken-and-egg problem of Simultaneous Localization and Mapping, a.k.a. *SLAM*. To this end, several types of mapping, localization, and simultaneous mapping and localization approaches were described. Later parts of this book will often refer to concepts from this chapter like localization and mapping (mostly independently). Thus, it is imperative that the readers make an attempt to understand the mapping and localization aspects thoroughly.

References

[1] M.J. Milford, Robot Navigation From Nature: Simultaneous Localisation, Mapping, and Path Planning Based on Hippocampal Models, vol. 41, Springer Science & Business Media, 2008.

[2] S. Thrun, et al., Robotic mapping: a survey, in: Exploring Artificial Intelligence in the New Millennium, 2002, pp. 1–35.

[3] S. Thrun, Learning metric-topological maps for indoor mobile robot navigation, Artificial Intelligence 99 (1) (1998) 21–71.

[4] A. Hornung, K.M. Wurm, M. Bennewitz, C. Stachniss, W. Burgard, OctoMap: an efficient probabilistic 3D mapping framework based on octrees, Autonomous Robots (2013), https://doi.org/10.1007/s10514-012-9321-0, Software available at http://octomap.github.com.

[5] F. Pasqualetti, A. Franchi, F. Bullo, On cooperative patrolling: optimal trajectories, complexity analysis, and approximation algorithms, IEEE Transactions on Robotics 28 (3) (2012) 592–606.

[6] N. Tomatis, I. Nourbakhsh, R. Siegwart, Combining topological and metric: a natural integration for simultaneous localization and map building, in: Proceedings of the Fourth European Workshop on Advanced Mobile Robots (Eurobot), ETH-Zürich, 2001.

[7] F. Dayoub, T. Morris, B. Upcroft, P. Corke, Vision-only autonomous navigation using topometric maps, in: Intelligent Robots and Systems (IROS), 2013 IEEE/RSJ International Conference on, IEEE, 2013, pp. 1923–1929.

[8] I. Kostavelis, A. Gasteratos, Semantic mapping for mobile robotics tasks: a survey, Robotics and Autonomous Systems 66 (2015) 86–103.

[9] A. Solin, M. Kok, N. Wahlström, T.B. Schön, S. Särkkä, Modeling and interpolation of the ambient magnetic field by Gaussian processes, IEEE Transactions on Robotics 34 (4) (2018) 1112–1127.

[10] P. Jensfelt, S. Kristensen, Active global localization for a mobile robot using multiple hypothesis tracking, IEEE Transactions on Robotics and Automation 17 (5) (2001) 748–760.

[11] E. Menegatti, M. Zoccarato, E. Pagello, H. Ishiguro, Image-based Monte Carlo localisation with omnidirectional images, Robotics and Autonomous Systems 48 (1) (2004) 17–30.

[12] F. Dellaert, D. Fox, W. Burgard, S. Thrun, Monte Carlo localization for mobile robots, in: Robotics and Automation, 1999. Proceedings. 1999 IEEE International Conference on, vol. 2, IEEE, 1999, pp. 1322–1328.

[13] V. Seshadri, G.V. Zaruba, M. Huber, A Bayesian sampling approach to in-door localization of wireless devices using received signal strength indication, in: Pervasive Computing and Communications, 2005. PerCom 2005. Third IEEE International Conference on, 2005, pp. 75–84.

[14] J. Ko, D. Fox, GP-BayesFilters: Bayesian filtering using Gaussian process prediction and observation models, Autonomous Robots 27 (1) (2009) 75–90.

[15] K.H. Low, N. Xu, J. Chen, K.K. Lim, E.B. Özgül, Generalized online sparse Gaussian processes with application to persistent mobile robot localization, in: Joint European Conference on Machine Learning and Knowledge Discovery in Databases, Springer, 2014, pp. 499–503.

[16] P. Jensfelt, H.I. Christensen, Pose tracking using laser scanning and minimalistic environmental models, IEEE Transactions on Robotics and Automation 17 (2) (2001) 138–147.

[17] B. Ristic, S. Arulampalam, N. Gordon, Beyond the Kalman Filter: Particle Filters for Tracking Applications, Artech House, 2003.

[18] S. Thrun, W. Burgard, D. Fox, Probabilistic Robotics, Massachusetts Institute of Technology, USA, 2005.
[19] H. Durrant-Whyte, T. Bailey, Simultaneous localization and mapping: part I, IEEE Robotics & Automation Magazine 13 (2) (2006) 99–110.
[20] T. Bailey, H. Durrant-Whyte, Simultaneous localization and mapping (SLAM): part II, IEEE Robotics & Automation Magazine 13 (3) (2006) 108–117.

Part II

The essentials
The building blocks

Contents

This part introduces the building blocks of this book to the readers in an incremental fashion. Special attention has been paid to keep the learning curve as smooth as possible to make this section and further reading amicable to readers from a variety of domains.

II.1 Preliminaries

In this chapter, basic mathematical tools involved in performing Bayesian inference are introduced and formalized.

II.2 Gaussian process (GP)

In this chapter, the keystone of this work, i.e., a non-parametric Bayesian inference method called Gaussian Process (GP) is introduced for first-time readers. This is a well known tool for performing non-parametric classification, regression, and optimization, and this chapter formalizes the mathematical notion of this model and discusses some of the potential applications and limitations of this model.

II.3 Coverage Path Planning (CPP)

This chapter discusses a class of path planning approaches that aim at maximizing area coverage called *Coverage Path Planning (CPP)*. Several existing path planners that belong to this category are discussed, along with their potential limitations if they were to be used for an environment monitoring task which is the core theme of this book.

II.4 Informative Path Planning (IPP)

Owing to the exhaustive exploration nature of the CPP framework, another class of path planners is investigated. Such planners are called *Informative Path Planning (IPP)* and focus on maximizing the amount of information as opposed to maximal area coverage.

Preliminaries
A primer

Give me 6 hours to cut down a tree and I will spend the first 4 hours to sharpen the axe.

Abraham Lincoln

Contents

Highlights

- Introduction to Bayesian Inference
- Refresher for multi-variate Gaussian distributions
- Primer on gradient descent
- Introduction to kernel trick
- Cholesky decomposition for avoiding explicit kernel inversion
- Utilizing the jitter for numerical stability

This chapter serves as a refresher for the basics of probability theory and Bayesian inference. Additionally, it mentions some optimization gimmicks for ensuring numerical stability and reducing computational and memory complexities when performing linear algebra operations or matrix manipulations. These gimmicks will come in handy in later parts of this book.

Multi-Robot Exploration for Environmental Monitoring. https://doi.org/10.1016/B978-0-12-817607-8.00018-6

5.1   Bayesian Inference (BI)

This section gives a brief overview of what is known as the *Bayesian Inference (BI)* [1]. Introduced by the English statistician Thomas Bayes in the early 1970s, it is a technique used to place probability distributions to reason (infer) about events. Since it is an inference mechanism that heavily relies on Bayes' Theorem, it got its name as Bayesian Inference. The concept is explained via an intuitive example as presented in Example 1 and more formally in Box 5.1.

Example 1 (Practical scenario). Consider a scenario where an imaginary customer named Bob wants to have a custom pair of jeans stitched exclusively for him. The problem to tackle here is:

How much raw material (denim) does he need to have his jeans stitched to his fitting?

To find an answer to this question, Bob must gather some information that could provide some hints for him. To address this concern, he speaks to his family tailor who advises that Bob should buy 1.5 meter of a 1.5 meter wide fabric. Now, this information serves as a *prior* information for when Bob sets out to buy raw material. However, Bob is very sceptical of the fact that if the raw material is not sufficient, all his investments (in terms of time, money, raw material, etc.) will all go to waste. So, he wants to be sure that he gets the correct size of denim cloth for his jeans. Now, Bob reaches out to his nearest garment supplier to acquire the requisite raw materials for his jeans. He informs the sales person that acquiring the exact dimensions is of utmost importance. He makes the purchase and returns home. Giving in to his scepticism, he wanted to ensure for himself that there were no errors made while measuring the piece of cloth.

Given his brief exposure to statistics, Bob knows that he needs to make a few measurements to make his results statistically significant. For the sake of illustration, let the length measurements be $\mathbf{y} = [1.49, 1.51, 1.48]$ meters for the 3 measurements that Bob took. Let y_{true} represent the true measurement that Bob expected to buy, i.e., 1.5 meter. Each of the $y_{i; i \in \{1,2,3\}}$ are independent and identically distributed (i.i.d.) and are assumed to follow an arbitrary Gaussian distribution such that $y_i \sim \mathcal{N}(y_{true}, 0.1)$. Then, the most likely measurement, also known as the Maximum Likelihood Estimate (MLE), is given by $y_{MLE} = \frac{1}{3} \sum_{i=1}^{3} y_i$. Thus, $y_{MLE} = 1.493^1$ meter.

While calculating the MLE, the prior information was not accounted for. However, this can easily be done if the prior information is converted into a *prior distribution*. This gives rise to $y_{true} \sim \mathcal{N}(1.5, 0.1)$. This again is a Gaussian distribution with the mean as the desired fabric length and the variance representing the expected margin for error. The interesting transformation leads from a true scalar value to a probability distribution over y_{true} which now acts a random variable. Intuitively, this represents

---

[1] Rounded to nearest 3 decimal places.

the belief over fabric length that Bob had, before actually making the 3 measurements.

Bob now wishes to calculate the posterior distribution over the data given that he knows the prior and likelihood models already (from above). In this context, this would mean updating the belief Bob has over the fabric length ($\mathcal{N}(1.5, 0.1)$) based on the likelihood model obtained from the 3 measurements Bob made. Thus, the posterior then becomes $P(\mathbf{y}|y_{true}) = \underset{y_{true}}{\arg\max} \dfrac{P(y_{true}|\mathbf{y}) P(\mathbf{y})}{P(y_{true})}$.

Box 5.1 Bayesian Inference (BI)

Prior
Prior refers to the information that is accrued before the empirical data acquisition. This helps formulate an initial belief for the occurrence of an event and is not dependent on the empirical data that will be acquired later in time. For some event A, the prior is represented by $P(A)$.

Likelihood
Likelihood as the name suggests defines the likelihood of occurrence of an event conditioned on another event that has already happened. For instance, consider two events A, B of which A has already happened. Then, the likelihood of B is defined conditioned on A as $P(B|A)$.

Posterior
Posterior refers to the updated belief distribution over the occurrence of an event once the empirical data has been observed, i.e., after the incorporation of new information. This can be done using Bayes Theorem which states that $P(A|B) = \dfrac{P(B|A)P(A)}{P(B)}$.

5.2 Multi-variate Gaussian (MVN)

When describing the probability distribution of a vector of random variables $\mathbf{x} = [x_1, \ldots, x_n]^T$, the probability distribution is said to be multi-variate normal, or multivariate Gaussian, if the density function is defined by

$$f(x) = \frac{1}{\sqrt{(2\pi)^n |\Sigma|}} \exp\left(-\frac{1}{2}(\mathbf{x} - \boldsymbol{\mu})^T \Sigma^{-1}(\mathbf{x} - \boldsymbol{\mu}),\right) \tag{5.1}$$

where $\boldsymbol{\mu} \in \mathbb{R}^n$ represents the mean vector and $\Sigma \in \mathbb{R}^{n \times n}$ represents the positive-definite covariance matrix. The mean represents the first moment of the distribution that explains the average or central tendency of the data. The covariance, on the other hand, encodes the second moment. The variance measures the spread of the observed samples about the mean; covariance encodes the same, only for pairs of observations. The term $(\mathbf{x} - \boldsymbol{\mu})^T \Sigma^{-1}(\mathbf{x} - \boldsymbol{\mu})$ represents the Mahalanobis squared distance [2] between the set of observations $\mathbf{x}$ and the mean $\boldsymbol{\mu}$ which, intuitively, can be interpreted as the measure of dissimilarity between the observations and the mean of the distribution.

5.3   Gradient Descent (GD)

Gradient descent (GD) is an optimization technique used widely in machine learning to optimize model parameters. It is an optimization method aimed exclusively at convex objective functions that iteratively suggests how to update the value of parameters.

5.3.1   What is a gradient?

Gradient measures the rate of change of output(s) as a function of changes in the input(s). In some settings, gradient is also called as the slope. For instance, consider the equation of a straight line passing through two points (x_1, y_1) and (x_2, y_2):

$$y_2 - y_1 = m(x_2 - x_1). \tag{5.2}$$

Here, m is called the slope or gradient of the line, which is illustrated by Fig. 5.1.

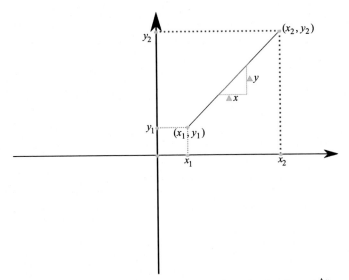

Figure 5.1 Illustration of a gradient via 2-point form of a line. Here, the slope is $m = \frac{\Delta y}{\Delta x}$.

5.3.2 How does it work?

Gradient descent is best illustrated via a toy example. Consider the scenario where a golfer wants to hit the ball into a marked goal in a minimal number of shots. For simplicity, consider only one hole in the golf field (i.e., no intermediate goals). This is shown in Fig. 5.2. This can be mathematically formulated as follows:

$$s^{[t+1]} \leftarrow s^{[t]} + \gamma \nabla(-f(s^{[t]})). \tag{5.3}$$

In the context of the example discussed above, in Eq. (5.3), $s^{[t]}$ represents the current location while $s^{[t+1]}$ represents the next location where the ball should land in order to reach the target as quickly as possible. The symbol γ represents the step-size $(\Delta_1, \ldots, \Delta_4)$ and $f(s^{[t]})$ is the cost/loss function being optimized. For the sake of discussion, let $f(s^{[t]})$ represent the mean squared distance between current ball location $(a, \ldots, e)$ and the target (y). The objective then is to minimize this distance as quickly as possible which is represented as a loss function in Fig. 5.2. This is exactly what gradient descent essentially tries to do- follow the gradient of the objective function in order to attain the minimum (or maximum, depending on the cost being optimized) as quickly as possible.

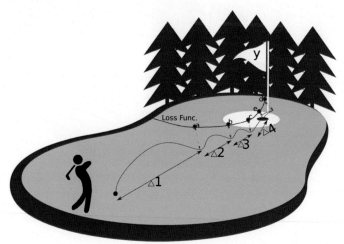

Figure 5.2 Illustrating gradient descent via a golf game example. Without any intermediate goals, the golfer tries to hit the ball into the marked hole (y) in as few shots as possible.

5.3.3 How optimal is the optimal solution?

In machine learning, Fig. 5.2 is the ideal case, as there is only one optimum where the loss function is at its lowest (shown by point e). This is what is referred to as the "global minimum". However, this may not always be the case. Consider a counter-example shown in Fig. 5.3, wherein there are two possible start configurations, where a golfer can start the game, and there are two different goals which the golfer can aim for. As the task is to hit the hole in the least number of trials, if the start position is on the left, the golfer should aim for hole y_2, otherwise, the golfer should aim for hole y_1. However, unaware of the fact that the reward associated with y_1 is 200 US$ while that of y_2 is 6000 US$, the golfer is most likely going to be "locally" optimal. What this entails is that (s)he can usually get the reward in the smallest number of trials but this need not necessarily be the highest of the rewards being offered. Gradient descent is prone to arriving at such local minima and failing to converge. Now, there are two important points when it comes to gradient descent, viz., initial guess and learning rate (γ) used in Eq. (5.3). To put things in perspective, this would mean that the golfer must be able to guess that since there are two holes, there is a chance that they have different rewards (maybe because the colors of flags are different or one is always easier to reach than the other). This would lead to an *initial guess* about the differential of the rewards which can then be optimized by carefully setting the parameter γ to increase the chances of reaching the global minimum.

Figure 5.3 Illustrating local minima problem with gradient descent via a golf game example. Here, assume that there are two holes with variable rewards. Depending on where the golfer starts from, (s)he would choose either of the holes and plan her/his shots accordingly. Since each golfer gets only one trial, (s)he would not know that there are variable rewards, and thus will be performing sub-optimally.

5.3.4   Types of gradient descent

The problem of "local" minima described above can be systematically addressed via a variety of gradient descent techniques available in the literature. The most commonly used methods are also described below.

5.3.4.1   Batch Gradient Descent (BGD)

Batch Gradient Descent (BGD), calculates the error for each training sample, but only after all the training examples have been evaluated, the model gets updated. This whole process is like a cycle and is referred to as a training epoch. The advantages (+) and disadvantages (−) of BGD are:

+ Computational efficiency
+ Stable convergence
− Convergence to local minima
− Entire training dataset must be available for each epoch

5.3.4.2 Stochastic Gradient Descent (SGD)

Stochastic gradient descent (SGD), in contrast to BGD, evaluates the error for each training example within the dataset. This means that it updates the parameters for each training example, one by one. The core strengths and weaknesses of SGD are:

+ Usually faster than BGD owing to sequential data processing
+ Frequent updates lead to detailed rate of improvement
− Frequent updates are computationally expensive (just like BGD which does the same in one shot)
− Frequency of updates induces noise in gradients which affects convergence

5.3.4.3 Mini-batch Gradient Descent (MBD)

Mini-batch Gradient Descent (MBD) lies in-between BGD and SGD. It simply splits the training dataset into small batches and performs an update for each of these batches. Therefore, it creates a balance between the robustness of SGD and the efficiency of BGD. The core strengths and weaknesses include:

+ Robust and efficient
− No rule-of-thumb for selecting mini-batch size

5.3.5 Gradient Descent with Random Restarts (GDR)

Gradient descent with random restarts (GDR) is a meta-algorithm designed on top of the gradient descent algorithm. Here, gradient descent is performed iteratively with each trial starting from a randomized initial condition (initial guess). The set of parameters that gives the maximal log-likelihood[2] are retained as the optimal set. This randomization procedure increases the chances of converging to a global optimum. However, this naturally begs the question: *How many random restarts would suffice?* This is an intriguing, yet challenging research question, given that the entire log-likelihood landscape cannot be known *a priori* owing to the complexity of the problem. However, preliminary findings have been reported in [3] and will not be investigated further within the scope of this book.

---

[2] Using the natural logarithmic scale is recommended for efficient computations and higher numerical precision.

5.3.6 Termination condition

The optimization is said to have terminated when any further changes in the values of the function parameters do not lead to significant changes in the cost, i.e., the output of the objective function being minimized has saturated. This is typically done by looking for small changes in error iteration-to-iteration (e.g., where the gradient is near zero).

5.4 Kernel trick

In order to perform non-linear regression using linear unbiased estimators, one would need to expand the feature set by including additional non-linear features. The complexity of the regressive model would then depend on the number and type of such additional features that need to be accounted for. In theory, the more such features are accounted for, the closer the regression curve gets to the true underlying function being modeled. However, as one might guess already, the computational and memory complexities increase exponentially for such models with the number of parameters being appended. To overcome this, the kernel trick is useful.

Kernel functions provide a way to manipulate data as though it were projected into a higher dimensional space, by operating on it in its original (low dimensional) space. This is illustrated via Fig. 5.4. First, consider the figure on the left which shows two classes of observations marked in red and green. Owing to the placement of the green sample, it is not possible to draw a straight line to segment the two classes, i.e., linear classification is not feasible. Using higher complexity classifiers can become computationally extensive owing to the complexity of the models, but projecting these samples onto a higher dimensional plane (shown on the right), using a polynomial transformation function, makes these samples separable by a hyper-plane. Thus, what was not linearly separable in $\mathcal{R}^2$ is now linearly separable in $\mathcal{R}^3$. Thus, on the right, the features would be expanded by additional features such that the new feature set contains x, y and $x^2 + y^2$ as used in this example. The high-dimensional mapped data is used to measure similarity (dot product between features). What the kernel trick essentially helps with is directly expressing the resultant dot product using the raw features in lower dimension. This does away with the need to map the features to a higher dimension explicitly by relying on kernels as defined in Definition 5.1.

Definition 5.1 (Kernel). Kernels are positive-definite matrices that represent a measure of similarity between inputs. For instance, consider two inputs, x_i and x_j, then $k(x_i, x_j)$ represents how similar are the observations of x_i and x_j as a function of the spatial separation of the inputs x_i, x_j.

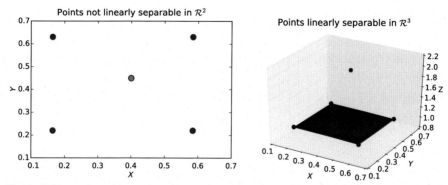

Figure 5.4 Illustrating the kernel trick via a non-linear classification example. On the left is a collection of two classes of observations marked in red and green, respectively. On the right is projection of features into higher dimensional space which makes them separable.

5.5 Cholesky decomposition for kernel inversion

Cholesky decomposition or factorization is a powerful numerical optimization technique that is widely used in linear algebra. It decomposes an Hermitian, positive definite matrix into a lower triangular and its conjugate component. These can later be used for optimally performing algebraic operations. One such operation that will be heavily used in this work will be the kernel inversion. In practice, inverting the kernel incurs the computational complexity of $\mathcal{O}(n^3)$ for a kernel of size $n \times n$. This grows exponentially as the size of kernel increases and this can be easily avoided by using Cholesky decomposition [4] instead, which has the computational complexity of $\mathcal{O}(n^3)$, but inverting the Cholesky factor only incurs $\mathcal{O}(n^2)$.

Example 5.5.1 (Cholesky decomposition). In this example, Cholesky decomposition will be used to solve a system of equations as opposed to direct matrix inversion which is computationally costlier as the size of matrix grows. Consider the following system of linear equations:

$$\begin{aligned}
x_1 - x_2 + 2x_3 &= 17, \\
-x_1 + 5x_2 - 4x_3 &= 31, \\
2x_1 - 4x_2 + 6x_3 &= -5.
\end{aligned} \tag{5.4}$$

In the matrix–vector notation, this can be written down as

$$\underbrace{\begin{bmatrix} 1 & -1 & 2 \\ -1 & 5 & -4 \\ 2 & -4 & 6 \end{bmatrix}}_{A} \times \underbrace{\begin{bmatrix} x_1 \\ x_2 \\ x_3 \end{bmatrix}}_{\vec{x}} = \underbrace{\begin{bmatrix} 17 \\ 31 \\ -5 \end{bmatrix}}_{\mathbf{b}}. \tag{5.5}$$

Now, a shorthand representation would be $Ax = \mathbf{b}$ where A represents the coefficient matrix, $\mathbf{x}$ represents the vector of variables, and $\mathbf{b}$ represents the vector of constants as also marked in Eq. (5.5). Notice that matrix A is symmetric positive-definite, and, hence, the Cholesky decomposition can be utilized. If this were not the case, then a more generic variant called LU decomposition could be used herewith. The only difference being, for Cholesky decomposition $U = L^T$.

Mathematically,

$$A\mathbf{x} = \mathbf{b} \equiv LU\mathbf{x} = \mathbf{b}. \tag{5.6}$$

Here, LU^3 refers to the L and U factor matrices of A. For this example,

$$L = \begin{bmatrix} 1 & 0 & 0 \\ -1 & 2 & 0 \\ 2 & -1 & 1 \end{bmatrix} \tag{5.7}$$

and

$$U = \begin{bmatrix} 1 & -1 & 2 \\ 0 & 2 & -1 \\ 0 & 0 & 1 \end{bmatrix}. \tag{5.8}$$

Then, the system of equations can be rewritten as

$$\underbrace{\underbrace{\begin{bmatrix} 1 & 0 & 0 \\ -1 & 2 & 0 \\ 2 & -1 & 1 \end{bmatrix}}_{L} \underbrace{\begin{bmatrix} 1 & -1 & 2 \\ 0 & 2 & -1 \\ 0 & 0 & 1 \end{bmatrix}}_{U}}_{A} \underbrace{\begin{bmatrix} x_1 \\ x_2 \\ x_3 \end{bmatrix}}_{\vec{\mathbf{x}}} = \underbrace{\begin{bmatrix} 17 \\ 31 \\ -5 \end{bmatrix}}_{\vec{\mathbf{b}}}. \tag{5.9}$$

Now, to solve the original system of equations, the intermediate solution to $LU\vec{\mathbf{x}} = \mathbf{b}$ needs to be found. For this, let $U\vec{\mathbf{x}} = \vec{\mathbf{y}}$ and solve $L\vec{\mathbf{y}} = \vec{\mathbf{b}}$. Once the solution $\vec{\mathbf{y}}$ is obtained, it can be substituted back to get the values of $\vec{\mathbf{x}}$. For this example, this becomes

$$\begin{bmatrix} 1 & 0 & 0 \\ -1 & 2 & 0 \\ 2 & -1 & 1 \end{bmatrix} \begin{bmatrix} y_1 \\ y_2 \\ y_3 \end{bmatrix} = \begin{bmatrix} 17 \\ 31 \\ -5 \end{bmatrix}, \tag{5.10}$$

solving which gives

$$\begin{bmatrix} y_1 \\ y_2 \\ y_3 \end{bmatrix} = \begin{bmatrix} 17 \\ 24 \\ -15 \end{bmatrix}. \tag{5.11}$$

This implies that

$$\begin{bmatrix} 1 & -1 & 2 \\ 0 & 2 & -1 \\ 0 & 0 & 1 \end{bmatrix} \begin{bmatrix} x_1 \\ x_2 \\ x_3 \end{bmatrix} = \begin{bmatrix} 17 \\ 24 \\ -15 \end{bmatrix}, \tag{5.12}$$

---

[3] This is a slight abuse of notation. This factorization is explicitly a Cholesky decomposition as $U = L^T$.

solving which finally returns the original unknown as

$$\begin{bmatrix} x_1 \\ x_2 \\ x_3 \end{bmatrix} = \begin{bmatrix} 51.5 \\ 4.5 \\ -15.0 \end{bmatrix}. \tag{5.13}$$

Owing to the sparsity of the upper and lower triangular factors, matrix manipulations are much more memory efficient. Also, note that $U = L^T$, and, thus, computationally only one factor needs to be computed and stored in memory. For the other factor, the previous one simply needs to be transposed. Like this, the need for actual matrix inversion can be by-passed in a computationally efficient and stable way.

5.6   Using jitter for numerical stability

When the entries of the rows of a covariance matrix are very similar, the matrix becomes ill-conditioned and inversion is also unstable. For understanding this, first the notion of the *condition number (con)* is defined, which can be used to evaluate how poorly conditioned the matrix is.

Definition 5.2 (Condition number (con)). Consider the covariance matrix where the entries of the first 2 rows are too similar as shown:

$$\begin{bmatrix} 1 & 0.9999 & 0 & 0 \\ 0.9999 & 1 & 0 & 0 \\ 0 & 0 & 1 & 0.1 \\ 0 & 0 & 0.1 & 1 \end{bmatrix}. \tag{5.14}$$

Then, $con = \dfrac{\lambda_{max}}{\lambda_{min}}$ where λ represents the eigenvalues. In this case, $con = 19999$, and the higher the value, the more ill-conditioned the matrix becomes.

Now, the conditioning problem can be addressed by adding a small positive quantity to the diagonal entries as shown:

$$\begin{bmatrix} 1.01 & 0.9999 & 0 & 0 \\ 0.9999 & 1.01 & 0 & 0 \\ 0 & 0 & 1.01 & 0.1 \\ 0 & 0 & 0.1 & 1.01 \end{bmatrix}. \tag{5.15}$$

For the revised matrix, $con = 199$, since the entries are now sufficiently dissimilar.

Remark. Jitter essentially is adding noise to the data, and hence, adding unnecessarily large noise to data can dilute the informativeness of the data. Thus, the jitter must always be kept sufficiently small to avoid numerical instabilities whilst retaining quality of the information to be processed.

5.7 Summary

This chapter summarized Bayesian Inference using a real-life example. Furthermore, a quick refresher on multi-variate Gaussian distributions, a.k.a. multi-variate normals, was given along with tools for inference, viz., gradient descent, its variants, and challenges faced when performing gradient based optimization of model parameters. Numerical optimization gimmicks like kernel trick, numerical jitter and Cholesky decomposition methods for numerically stable matrix manipulation were discussed. The next chapter scales up multi-variate Gaussian distribution to incorporate infinitely many random variables. This gives rise to what is now known as Gaussian Process, a widely accepted non-parametric Bayesian method that can be used for classification, regression, and optimization problems alike.

References

[1] G.E. Box, G.C. Tiao, Bayesian Inference in Statistical Analysis, vol. 40, John Wiley & Sons, 2011.
[2] G.J. McLachlan, Mahalanobis distance, Resonance 4 (6) (1999) 20–26.
[3] T. Dick, E. Wong, C. Dann, How Many Random Restarts Are Enough?, Google Scholar, 2014.
[4] W.H. Press, S.A. Teukolsky, W.T. Vetterling, B.P. Flannery, Numerical recipes 3rd edition: the art of scientific computing, Cambridge University Press, 2007.

Gaussian process
The function space view

6

Normal is the wrong name often used for average.

Henry S. Haskins

Contents

Highlights

- Formalizing Gaussian processes (GPs)
- Kernel jargon
- Inference with GPs
- Limitations of conventional GPs

Multi-Robot Exploration for Environmental Monitoring. https://doi.org/10.1016/B978-0-12-817607-8.00019-8

- Approximate GPs for computational efficiency
- Applications of GPs
- Hands-on experience with GPs

This chapter serves as an introduction to a non-parametric Bayesian method called Gaussian Process (GP). Whilst formalizing the computational form of GPs, the chapter also highlights a variety of jargons that exist in the literature associated with GPs. This chapter primarily discusses the regression capabilities of GPs, although they can be very well used for other applications like classification and optimization. The aim here is to describe both the strengths and weaknesses of GPs and discuss why the GPs are a common model of choice of complex regression problems.

6.1   Gaussian Process (GP)

Gaussian Processes (GPs) are non-parametric Bayesian inference methods that utilize Gaussian distributions [1]. Owing to their Gaussian nature, they can be fully described in terms of the mean and covariance functions.

The term "non-parametric" is a misnomer often leading the readers to believe that such methods are free of parameters altogether. Note that this is a severe misconception, and readers are hereby cautioned to read the following parts of the chapter very carefully.

In the rest of this chapter, first the notion of parametric vs. non-parametric will be clarified via an intuitive example. This will be followed by notational conventions which will be used for formalizing the mathematical notion of GPs. In the later parts of this chapter, inference mechanisms for training the GPs on observed data, limitations of conventional GPs and approximation methods are also introduced. This terminology will be frequently referred to, throughout this book, so it is imperative to understand it thoroughly before reading on.

6.1.1   Parametric versus non-parametric

GPs are non-parametric methods but this term can be severely misleading. In an attempt to avoid confusing notions of the phrase "non-parametric", here, an example is provided to clearly define the setting.

Parametric methods, as the name suggests, use parameters to define the functional forms. Non-parametric, on the other hand, is a misnomer. It does not mean that these methods are parameter-free. It essentially means that such methods do not make assumptions about the functional form of the model being considered. For a deeper insight, consider the example illustrated in Box 6.2 which represents the dilemma of how complex the model should be, to correctly estimate the number of consumers for a catering scenario. This dilemma occurs owing to the reliance on the *parametric* form of solution, i.e., the consumer company managers want to restrain their capabilities by stating how many degrees of freedom the model can have. As opposed to this, in a *non-parametric setting*, this can easily be solved by fitting all possible functions of varying complexities, and then selecting those which are least complex, yet perform well in fitting the data.

Box 6.2 Parametric vs non-parametric formulations

Consider the following scenario: A catering company ABC serves at a university canteen and is just getting started. They want to be sure that all staff and students are well catered and there is always sufficient food. An overview of this setting is shown in Fig. 6.1.

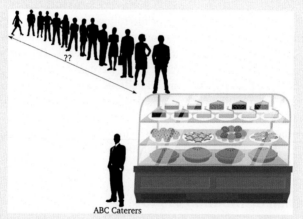

ABC Caterers

Figure 6.1 The catering scenario for ABC Caterers.

Given the power of machine learning models, they want to use them to base their decisions on the data instead of just using wild-guessing techniques. For this, they would need some observations (number of consumers on a daily basis) to fit a model. Let these observations be the represented by a red dot (●) as shown in Fig. 6.2. The observations here represent the number of consumers that were served on any given day. Assume that three such data points were logged with the number of consumers on day 2 significantly higher than on either of the other days. Now, the catering company needs to decide how many consumers to expect for the upcoming days.

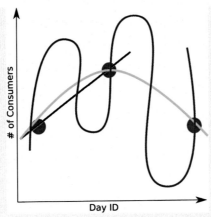

Figure 6.2 Illustration of the difference between parametric and non-parametric setting using a catering scenario.

Now, this can be addressed in one of the following ways: the catering company can rely on simple linear regression models and, say, a straight line (represented in black) should suffice to represent the trends. But very soon the company realizes that it is not enough to fit the data so that it increases the complexity of the model by choosing a quadratic model (represented by green curve). All was good until someone asked what would happen if they were to use an even more complex model (represented by the blue curve). Maybe it can capture even more data as it comes along? This dilemma is addressed using the non-parametric Bayesian methods as described next.

6.1.2   Gaussian distribution vs process

Gaussian probability distributions like the multi-variate normal (MVN) as discussed in Chap. 5 place distributions over vectors of random variables such that the "distribution" is completely defined by the **mean vector** and the **covariance matrix**. As opposed to this, a Gaussian "process" places distributions over functions such that the process is now defined in terms of **mean function** and **covariance function**. Placing distribution over functions provides the flexibility to consider infinitely many random variables. However, one would never need to consider this infinite-dimensional object as a whole in a real application. The reason for this is primarily that no matter how large a dataset is acquired, it is still finite, and no matter where in the domain predictions are to be made, they also remain finite (owing to computational limitations). Under this setting, any finite subset of the random variables when modeled as a GP primarily reduces to joint Gaussian, which depending on the dimensionality of the input space, is representable by the MVN probability distribution function (pdf) as presented earlier.

6.1.3 Notational conventions

Just like conventional Gaussian distributions, the mean function is represented by μ while the noise-free covariance (kernel) is represented by $\mathcal{K}$. As most of the real-world data is noisy owing to sensor noise or environmental disturbances, etc., noise is also added to kernels which are then represented by $\mathcal{K}_\epsilon$ to account for i.i.d. white noise. For the inputs, let $\mathbf{x}^-$ represent the list of observed inputs and $\mathbf{y}^-$ be the list of corresponding targets/observations. Similarly, let $\mathbf{x}^*$ represent the set of unobserved inputs for which the predictions $(\mathbf{y}^*)$ need to be obtained as posterior. Also, let f be the underlying function that is to be approximated by using a GP via the data (D) is denoted in terms of the observation tuples as $D = \langle \mathbf{x}^-, \mathbf{y}^- \rangle$.

6.1.4 Mean function

For convenience, without any loss of generality, most researchers set the prior mean function to be zero [1]. This assumption is not restrictive as the posterior mean takes the conjugate form of the likelihood. However, if the prior mean function is known to be non-zero, a change of variables proves helpful such that, if $f \sim \mathcal{GP}(\mu, \mathcal{K}(\cdot, \cdot))$, then $f' \triangleq f - \mu$ is the new zero mean GP with $f' \sim \mathcal{GP}(0, \mathcal{K})$. Hence, instead of doing inference for f, the same can be done for f', and once the posterior mean is obtained, it can simply be added back to the prior mean (μ) to obtain the posterior for f.

6.1.5 Covariance function

GPs model the correlations between observations (or targets) as a function of the inputs by utilizing the kernel trick. This is why GPs are considered as supervised learning methods since the apt choice of kernel has a significant bearing on the quality of the learnt model and this relies on some prior information about how the data varies with respect to the domain. Further details of the well-established kernels from the literature are summarized below. Consider any pair of targets $y, y' \in \mathbf{y}^-$, then the covariance between them can be denoted by $cov(y, y') = \sigma(y, y')$. Iterating over all such pairs, i.e., $\forall y, y' \in \mathbf{y}^-$, a covariance matrix $\mathcal{K}$ is obtained.

6.1.6 Choices of kernels

A multitude of covariance functions exist as explained in [2,3] and furthermore can be generated by combinations of existing kernels as explained in [4, Chap. 2]. However, for the sake of completeness of this book, some of the most widely used kernels are summarized below [5].

In Table 6.1, $d \triangleq (\mathbf{x} - \mathbf{x}')^T L^{-1}(\mathbf{x} - \mathbf{x}')$ and $\tilde{\mathbf{x}} \triangleq [1, \mathbf{x}]$.

Table 6.1 Some of the well-established kernel functions.

Name	Description $\mathcal{K}(\mathbf{x}, \mathbf{x'})$	Hyper-parameters
Constant	σ_o^2	$\boldsymbol{\theta} \triangleq \{\sigma_o\}$
Linear	$\sigma_{sig}^2 (\sigma_o^2 + \mathbf{x}^T L \mathbf{x'})$	$\boldsymbol{\theta} \triangleq \{\sigma_{sig}, \sigma_o, L\}$
Gaussian Noise	$\sigma_{sig}^2 \delta_{\mathbf{x}, \mathbf{x'}}$	$\boldsymbol{\theta} \triangleq \{\sigma_{sig}^2\}$
Exponential	$\sigma_{sig}^2 \exp\left(-\sqrt{\mathbf{d}}\right)$	$\boldsymbol{\theta} \triangleq \{\sigma_{sig}, L\}$
γ-Exponential	$\sigma_{sig}^2 \exp\left(-\mathbf{d}^{\gamma/2}\right)$	$\boldsymbol{\theta} \triangleq \{\sigma_{sig}, L, \gamma\}$
Squared Exponential (RBF) with ARD	$\sigma_{sig}^2 \exp\left(-\frac{1}{2}(\mathbf{d})\right)$	$\boldsymbol{\theta} \triangleq \{\sigma_{sig}, L\}$
Squared Exponential (RBF) without ARD	$\sigma_{sig}^2 \exp\left(-\frac{1}{2l^2}(\mathbf{x} - \mathbf{x'})^T (\mathbf{x} - \mathbf{x'})\right)$	$\boldsymbol{\theta} \triangleq \{\sigma_{sig}, l\}$
Matern $\frac{1}{2}$	$\sigma_{sig}^2 \exp\left(-\mathbf{d}\right)$	$\boldsymbol{\theta} \triangleq \{\sigma_{sig}, L\}$
Matern $\frac{3}{2}$	$\sigma_{sig}^2 (1 + \sqrt{3\mathbf{d}}) \exp(-\sqrt{3\mathbf{d}})$	$\boldsymbol{\theta} \triangleq \{\sigma_{sig}, L\}$
Matern $\frac{5}{2}$	$\sigma_{sig}^2 (1 + \sqrt{5\mathbf{d}} + \frac{5\mathbf{d}}{3}) \exp(-\sqrt{5\mathbf{d}})$	$\boldsymbol{\theta} \triangleq \{\sigma_{sig}, L\}$
Rational Quadratic	$\sigma_{sig}^2 \left(1 + \frac{\mathbf{d}}{2\alpha}\right)^{-\alpha}$	$\boldsymbol{\theta} \triangleq \{\sigma_{sig}, L, \alpha\}$
Polynomial	$\sigma_{sig}^2 (\sigma_o^2 + \mathbf{x}^T L \mathbf{x'})^p$	$\boldsymbol{\theta} \triangleq \{\sigma_{sig}, \sigma_o, L, p\}$
Periodic Kernel	$\sigma_{sig}^2 \exp\left(\frac{2\sin^2(\pi T \sqrt{\mathbf{d}})}{\rho^2}\right)$	$\boldsymbol{\theta} \triangleq \{\sigma_{sig}, L, T, \rho\}$
Neural Network	$\frac{2}{\pi} \sin^{-1}\left(\frac{2\tilde{\mathbf{x}}^T \Sigma \tilde{\mathbf{x}}'}{\sqrt{(1 + 2\tilde{\mathbf{x}}^T \Sigma \tilde{\mathbf{x}})(1 + 2\tilde{\mathbf{x}}'^T \Sigma \tilde{\mathbf{x}}')}}\right)$	$\boldsymbol{\theta} \triangleq \{\Sigma\}$

6.2 Kernel jargons

Various classes of kernel functions are known in the literature and are usually addressed with specific terms (jargons). For the ease of the readers, they are categorized and detailed below along with an intuitive notion of the purpose served by the kernel itself.

- **Kernels**
  Kernels are a flexible way to represent the data so that it can be used to compare the samples in a complex space. A kernel is basically a non-linear non-parametric transformation function that maps the data to a higher dimensional space enabling linearity and easier comparisons of complex features.[1] By doing this mapping, kernels enable the machine learning approaches to utilize a linear model in the new input space to quantify similarity between a pair of objects x and x'. This is equivalent to regression by utilizing infinitely many Gaussian shaped basis functions placed everywhere and not just over the training points.

- **Stationary vs. non-stationary kernels**
  A translation-invariant kernel is called stationary. This means that for any two inputs x, x', the following holds:

$$\mathcal{K}(x, x') = \mathcal{K}(x + \Delta, x' + \Delta). \tag{6.1}$$

From Eq. (6.1), it can be intuitively inferred that a stationary kernel is a kernel that is purely a function of $||x - x'||$ and not the actual value of x, thereby making it translation invariant. Furthermore, this also means that the kernel is rotation invariant, so that the kernel is said to be an *isotropic* stationary kernel, and, conversely, if the separation in features is a function of direction, then such a kernel is termed as *anisotropic* stationary kernel [6].

It is often the case that the correlations do not vary uniformly throughout the entire domain. Thus, generic kernels that can explain the underlying dynamics as a function of the input itself are required. This means that

$$\mathcal{K}(x, x') = \int_{\mathcal{R}^d} \mathcal{K}_x(\cdot)\mathcal{K}_{x'}(\cdot)d(\cdot). \tag{6.2}$$

The kernels of the form given by Eq. (6.2) are referred to as non-stationary kernels since the correlation varies with the input. Works like [7] discuss a novel non-parametric inference method when dealing with non-stationary inputs which otherwise renders an intractable posterior.

- **Separable vs. non-separable kernels**
  Usually when dealing with inputs represented by the spatio-temporal tuple as $[\mathbf{x} \in \mathcal{R}^2, t \in \mathcal{R}]$, a natural way to build kernels dealing with spatial and tempo-

---

[1] For a refresher the readers can turn to Sect. 5.4 of Chap. 5.

ral domains is to multiply a spatial kernel with a temporal kernel as

$$\mathcal{K}_{st}((\mathbf{x}, \mathbf{x}'), (t, t')) = \underbrace{\mathcal{K}_s(\mathbf{x}, \mathbf{x}')}_{\text{spatial kernel}} \cdot \underbrace{\mathcal{K}_t(t, t')}_{\text{temporal kernel}} . \tag{6.3}$$

This procedure is deemed valid and renders a valid kernel only if the kernels/domains are separable. However, if a kernel is not reducible into spatial and temporal domains as explained in Eq. (6.3), then works like [8] can be utilized to create non-separable non-stationary space–time kernels that even allow for distinct variations in either domain. To test for separability, Hilbert–Schmidt Independence Criterion (HSIC) [9] can be used.

- **Homoscedastic vs. Heteroscedastic Kernels (Inference with Noisy Data)**
  Usually the observations being gathered for training are noisy owing to sensor modalities utilized to gather the data. To account for such factors, a noise term is additionally added to the covariance function. For any two inputs x, x', this can be done as follows:

$$\mathcal{K}_\epsilon(x, x') = \mathcal{K}(x, x') + \delta\sigma_n^2$$
$$\Rightarrow \mathcal{K}_\epsilon = \mathcal{K} + \sigma_n^2\mathbb{I}. \tag{6.4}$$

In Eq. (6.4), the symbol δ refers to the Dirac-delta function which is used to check if $x = x'$ and σ_n represents the distribution of the noise. Hence, the nature of noise is a function of the data being observed. If $\sigma_n \sim \mathcal{N}(0, \epsilon)$, then a uniform noise is assumed for the entire domain and this is termed as *homoscedasticity*. On the other hand, if $\sigma_n \sim \mathcal{N}(0, \epsilon(x))$, then, input-dependent noise model is obtained and is termed as *heteroscedasticity*.

- **Hyper-parameters**
  Also known as the free parameters, these hyper-parameters control the behavior of the GP. Intuitively, the characteristic length scalesrepresent the distance one must move in the input space before the function value can change *significantly*. Thus, short length-scales mean the error bars (i.e., predictive variance) can grow rapidly away from the data points whilst large length-scales imply irrelevant features (function value would be constant function of that feature input). When dealing with multi-dimensional input spaces, it is possible that the variation is not similar for each dimension and hence, the method of *Automatic Relevance Determination (ARD)* can be deployed which has been so named since the model determines the "relevance" of length scales per dimension. Besides the characteristic length scales, for convenience, two additional parameters are also accounted for in the same category, viz., signal variance which defines the amplitude of variance in the signal being monitored and similarly noise variance which captures the noise amplitude. Together, the entire set of signal variance, characteristic length scales, and noise variance are referred to as hyper-parameters for the scope of this work.

6.3 Bayesian inference

As was stated earlier, GPs are non-parametric "Bayesian" methods. This means that there should be a prior and a likelihood which give rise to the posterior and this should be updated as more data is observed. These terminologies were previously introduced in Chap. 5.

6.3.1 Prior

As the GPs are defined in terms of mean and covariance functions, prior mean and prior covariance need to be specified (either manually or by model selection). In practice, most researchers prefer to set the prior mean to zero [1], without any loss of generality. However, if non-zero prior means are desired, a simple change of variables can be utilized as explained earlier. For this book, in keeping with the common research practices, zero mean priors were assumed.

As covariance kernels define the correlations between observations, they either need to be pre-selected by the human or automatically selected using model selection criterion. When manually setting the functional form of the covariance from one of the well-established kernels or their combinations thereof, one gets what is known as the prior kernel. This kernel has some *a priori* chosen hyper-parameters which are then tuned based on the observations accrued. However, for some applications this choice may be non-trivial and, hence, model selection methods rely on approximate set coding [10] or approximate Bayesian computation (ABC) [11]. Throughout this book, manually selected priors will be used, which were chosen based on the performance reported by peers who used similar kernels.

6.3.2 Likelihood

In Chap. 5, the following relationship was introduced:

$$\text{Posterior} \propto \text{Likelihood} \times \text{Prior}. \tag{6.5}$$

While Sect. 6.3.1 defines the prior mean and covariance functions, a likelihood function is yet to be defined before posterior can be obtained. Likelihood, as the name suggests, explains how likely it is to observe the data given the current model, i.e., given the current hyper-parameters of the kernel, how likely it is to see the data. The higher the likelihood, the better is the model in explaining the data.

For this, the probability density function of a multi-variate Gaussian (Eq. (5.1)) will be used. From this equation, the log-likelihood can be obtained as follows:

$$\mathcal{LL} = \underbrace{-\frac{n}{2}\log(2\pi) - \frac{n}{2}\log(\sigma_{sig}{}^2)}_{\text{Const.}} \underbrace{- \frac{1}{2}\log|\mathcal{K}|}_{\text{Complexity}} \underbrace{- \frac{1}{2\sigma_{sig}{}^2}\mathbf{y}^T\mathcal{K}^{-1}\mathbf{y}}_{\text{Data fit}}. \tag{6.6}$$

In Eq. (6.6), σ_{sig}^2 represents the signal variance and the equation penalizes very complicated models (Occam's razor principle [12]) whilst accounting for good model fit; $\mathbf{y} \in \mathbf{y}^-$ represents the list of observations/targets that were observed.

6.3.3 Posterior

Posterior, as the name suggests, is used for predictions. Mathematically, this looks like:

$$
\mu_{f|D} \triangleq \underbrace{\mu^{*0}}_{\text{Prior}} + \underbrace{\mathcal{K}^{*T}\mathcal{K}^{-1}(\mathbf{f}-\mu)}_{\text{Corrector}},
$$
$$
\mathcal{K}_{f|D} \triangleq \underbrace{\underbrace{\mathcal{K}^{**}}_{\text{Prior}} - \underbrace{\mathcal{K}^{*T}\mathcal{K}^{-1}\mathcal{K}^{*}}_{\text{Evidence}}}_{\text{Reduction in Variance}}, \tag{6.7}
$$

where $\mu_{f|D}$ refers to the posterior mean which represents the predicted observations over unobserved inputs and $\mathcal{K}_{f|D}$ represents the posterior covariance which represents a measure of confidence over the predictions being made. Further, $\mathcal{K}$, $\mathcal{K}^*$, and $\mathcal{K}^{**}$ are short-hand notations for the following covariances: $\mathcal{K} = \mathcal{K}(\mathbf{x}^-, \mathbf{x}^-)$, $\mathcal{K}^* = \mathcal{K}(\mathbf{x}^*, \mathbf{x}^-)$, and $\mathcal{K}^{**} = \mathcal{K}(\mathbf{x}^*, \mathbf{x}^*)$. Similarly, the prior mean is given by $\mu^* = \mu(\mathbf{x}^*)$ over the unobserved inputs.

6.3.4 Maximum Likelihood Estimation (MLE)

With the log-likelihood defined by Eq. (6.6), the following hyper-parameters need to be inferred: $\theta = \{\sigma_{sig}, l, \sigma_n\}$. This can be done by iteratively maximizing the log-likelihood by updating the parameters in the steepest direction of the gradient of log-likelihood with respect to the parameter being optimized. Here, the log-likelihood is being used as the (natural) logarithm is a monotonically increasing function and is numerically quite stable. The gradient of Eq. (6.6) is given by:

$$
\frac{\partial}{\partial \theta_j}(\mathcal{LL}) = \frac{1}{2}\mathbf{y}^T\mathcal{K}^{-1}\mathbf{y}\frac{\partial \mathcal{K}}{\partial \theta_j}\mathcal{K}^{-1}\mathbf{y} - \frac{1}{2}\text{tr}\left(\frac{\partial \mathcal{K}}{\partial \theta}\right), \tag{6.8}
$$

where $\text{tr}(\cdot)$ represents the trace of the matrix comprised of the entries given by $\left(\frac{\partial \mathcal{K}}{\partial \theta}\right)$.

6.4 Entropy

The core strength of GPs lies not only in the fact that they are kernel-based methods that can elegantly model the correlations in observations but also that they provide analytical solutions for generating posteriors. Additionally, these solutions come with a measure of uncertainty of the model over the domain which can be defined in terms of the "entropy". Let $\mathbb{H}$ represent the entropy of the GP model generated (i.e., the posterior), then the entropy can be formally defined as

$$\mathbb{H} \triangleq \text{tr}[\mathcal{K}_{f|D}], \tag{6.9}$$

or,

$$\mathbb{H} \triangleq \frac{|\#\mathcal{K}_{f|D}|}{2} \log(2\pi e) + \frac{1}{2} \log(|\mathcal{K}_{f|D}|). \tag{6.10}$$

In Eq. (6.9), conditional independence between targets is assumed while this is aptly accounted for by Eq. (6.10) [13]. Within the scope of this book, wherever entropy is referred to, Eq. (6.10) will be utilized as it helps avoid model over-fitting (over-estimation of variance).

6.5 Multi-output GPs (MOGPs)

Thus far, single output GPs were presented. This is evident from the posterior Eq. (6.7) where the posterior mean and covariance represented the belief over the target values of only one type at a variety of unobserved query points. In [14], multi-output GPs were discussed. The multi-output Gaussian Process (MOGP) represents a setting where multiple interacting outputs are to be modeled, which makes the inference very challenging, as the positive definite nature of the kernel may be hard to maintain if not aptly handled. To this end, [15] describes a novel iterative formulation of a multi-output Gaussian process that can build and exploit a probabilistic model of the environmental variables being measured (including the correlations and delays that exist between them). In [16], a novel extension of MOGP is discussed, which deals with multiple outputs that are heterogeneous in nature. Aside from learning spatio-temporal correlations across multiple outputs to guarantee a positive definite kernel, MOGPs can also be used as a tool for quantifying uncertainty [17].

6.6 Limitations of GPs

GPs are beset with several limitations. Some of them are summarized below.

- **Slow Inference.** Optimizing the hyper-parameters and predicting the function values at locations of interest incurs $\mathcal{O}(n)^3$ time complexity and $\mathcal{O}(n)^2$ memory complexity, with n being the size of the dataset being modeled. This restricts the exact inference only to the order of 10^3 data points. Any larger datasets must be addressed using approximate inference methods.
- **Apt Kernel Choice.** Although GPs are non-parametric in nature, the covariance function needs to be chosen in accordance with the dynamics of the field being modeled since the kernel essentially represents the input space. Optimizing the marginal log-likelihood from Eq. (6.6) can provide an optimal parameter setting but the supervisor needs to set the parametric form of the kernel.

- **No consideration for robot resources.** GPs were designed as machine learning models that can reach high computational speeds using GPU optimization, but when deploying real robots to serve as mobile sensor nodes, relaying the data to a powerful base node, approximate methods need to be used to optimize the GPs to work with limited hardware capabilities.
- **Designed for batch processing.** GPs were inherently designed for batch-processing and big-data applications. Optimizing the parameters using a stream of sequential data is a rather early-stage research domain.

Having discussed both the pros and cons of using the GPs and several potential applications where they can be seamlessly integrated, one daunting question still remains: *What makes them the model of choice?* The answer to this question is rather concise. The fact that these models have a flexible covariance kernel that can be used to explain the correlations in most of the potential applications discussed above along with the ability to quantity the predictive capabilities makes them the model of choice. Other machine learning methods like Neural Networks, Gaussian Mixture Models, etc., could be used but then additional models would be required to quantify their performance. Thus, for all intents and purposes, this work will utilize GPs as the model that can effectively explain the underlying dynamics of the phenomenon of interest, viz., for *Intelligent Environment Monitoring*.

6.7 Approximate GPs

The main bottleneck of GPs is the $\mathcal{O}(n)^3$ computational complexity and $\mathcal{O}(n)^2$ memory complexity for inverting a kernel with n data points. The most intuitive fix to this problem would then be to figure out a way to reduce the size of n without compromising the performance. According to the survey [18], there are two ways of scaling the GPs to accommodate big-data, as discussed below.

6.7.1 Global approximation methods

In these methods, global distillation of the training data is carried out such that only a smaller subset of the training data is retained and used for inference. The mechanisms for cherry-picking the dataset subset are given below.

6.7.1.1 Random subset selection

As explained in [19], a possibility for *passively* selecting a subset of the training data is by selecting a random subset of the overall training data. In essence, if the size of random subset is aptly chosen, it will be much smaller than the original training set thereby aiding the bottleneck of computational complexity.

6.7.1.2   Active subset selection

As opposed to *passively* selecting the data, differential entropy score [20] can be used for subset selection as an *active* selection mechanism. This method is known to perform well [21] and also takes into account the quality of the data that will later be used for training and inference of GPs.

6.7.1.3   Sparsifying the kernel

Another approach could be to sparsify the kernel itself. This involves removing the un-important data samples whose correlations are less than a preset threshold. By doing so, several zero entries are introduced in the kernel, making it sparse, and thus, light on memory and computational costs.

6.7.1.4   Sparse approximation of the kernel

As opposed to sparsifying the kernel by thresholding which results in the loss of correlations, low-rank representation (Nyström approximation) of the kernel can be investigated. This is done by selecting a set of pseudo-inputs which are not necessarily sampled from the original dataset. As the size m of this dataset is selected by the user, the computational complexity is now $\mathcal{O}(nm)^2$ and memory complexity is $\mathcal{O}(nm)$ while $m \ll n$.

6.7.1.5   Variational sparse approximation

Variational sparse approximation utilizes variational inference (VI) [22] which is a machine learning technique which attempts at modelling the otherwise hard to model probability densities. This is done by utilizing a family of densities and then selecting the member which is closest to the target density.

As an example, consider a sales personnel pitching an soon-to-be launched product to a potential client. The client suddenly asks a question which the sales personnel did not anticipate. Given the high stakes, the sales personnel poses an alternative variant to the same question and eventually ends up answering the reformulated question as opposed to answering the original question. This is the notion of VI.

The "closeness" between the target and the approximate distributions in the case of GPs is measured in terms of the Kullback–Leibler divergence (KLD) [23] between densities. This idea applied for GP regression was first presented in [24] and was further enhanced in [25] to maintain linear complexity.

6.7.2   Local approximation methods

As opposed to global approximation methods which select a representative subset of the entire training data, local methods deploy the divide-and-conquer mechanism. Here, the strategy is to slice the entire data into multiple subsets and then train an ensemble (family of GPs acting as a whole) over the subsets to generate a smooth posterior. One such model called the partitioned variational inference (PVI) is explained in [26] for continuous learning. Another alternative approach was proposed in [27] for scaling GPs to harness the multi-core computer architectures for fast sequential updates.

6.8 Applications of GPs

GPs are highly modular non-parametric Bayesian methods and their flexibility in the choice of kernels allows their application to *classification*, *optimization*, and *regression* applications alike. Some potential applications from each of these aspects are summarized below.

6.8.1 GPs applied to classifications tasks

- Usually when a patient is newly admitted to a hospital, the staff record several vital statistics like age, blood pressure, height, weight, previous medical ailments, current medications (if any), allergies to medications (if any), nature of ailment, etc. Based on these parameters, a decision needs to be made if a patient is to be admitted to the general ward (GW) or the intensive care unit (ICU). Thus, there are two class labels, viz., GW, ICU, and based on a dozen or so variables (vital stats) an accurate decision needs to be made. The problem could be further assessed as scheduling problem such that given the critical nature of the patients, the patients are assigned to a queue and scheduled to attend in their respective wards.
- When applying for credit cards or bank loans, the concerned agencies receive several applications containing information about the applicants age, marital status, annual income, previous deficits, type of card/loan, etc. Based on these parameters, a decision needs to be made if the applicant can be selected or rejected. Some applicants, however, do not fit into either of these class labels and fall into a gray zone which must be manually handled by the supervisor or can be accounted for by adjusting the classification threshold of the classifier.
- Several applicants join the motor driver school to learn how to drive 2/4 wheeled vehicles either in low weight (motorbikes), moderate weight (cars), or heavy weight (trucks/buses) category. While evaluating the applicants, not only the theoretical test and road tests are important where the applicant is graded based on his/her knowledge of road signs and attentiveness while driving, but also a stern background check is essential. The applicant should be in the appropriate driving age limits, physically fit with optimal eye-sight, and should not have any prior offenses on record. After accounting for all such parameters, a decision can be made to allow or revoke the application.
- During an auction, several bidders place a bid for the items on display but depending on the age, background, financial status, etc., of the attendees a suitable starting bid needs to be selected.
- When a large industrial plant is set up, before a decision is made to buy a certain patch of land, several aspects need to be accounted for, e.g., proximity to the re-

gions that will be used to acquire the raw materials, proximity to the buyers so that transport cost can be minimized, size of the work force, chances of attracting the said size of work force, etc. All these parameters should be evaluated for the most feasible combination, only then can a decision be made if the land under consideration should be purchased or not.

- Space exploration missions are carried out using either mobile robots or cameras which send noisy images of sub-optimal (owing to bandwidth limitations) resolution to the base station. Scientists then need to label all the objects in the scene as stars, galaxy, craters, etc., with a very high certainty.
- Handwritten digit recognition has been extensively studied and has intrigued researchers for quite some time. Classifying the acquired digit as one of $\{0, \dots, 9\}$ can also be assisted by supervised learning using GPs.
- Occupancy Grid Maps (OGMs) have also been developed by utilizing GPs to observe the collection of laser beams that bounce off the environment in order to detect if a location is *free* or *occupied*.

6.8.2   GPs applied to regression tasks

- Inverse Kinematics of robot arms are used to deduce the joint angles which must be set to reach a required end-effector position.
- For soil mapping, erosion mapping, surface water monitoring-like environmental applications based on the training samples gathered by a robot or static sensors, a model is generated using GPs to interpolate and generate predictions over other unforeseen regions.
- In financial mathematics, GPs can be used to predict stock market and urban housing prices based on the trends depicted in the past.
- Just like localization w.r.t. the geometric configuration of the environment, localization can also be performed with respect to the measurement domain. This is alternatively known as *signal strength based localization* where the robot needs to infer its location based on currently acquired noisy observations.

The application of interest, as far as this work is concerned, is *regression* and in particular *Environment Monitoring* which some people also refer to as *Intelligent Environment Monitoring* [28], since robots are required to intelligently select and observe parts of a large-scale phenomenon.

6.8.3   GP applied to Bayesian Optimization (BO)

Besides non-linear classification and regressions applications as mentioned above, another powerful application of GPs in the machine learning literature is in the field of Bayesian optimization. The intuition is that the function f being modeled by the GPs is unknown and sampling from it is costly. For example, let $\mathbf{x}$ represent the geographic location(s) and $f(\mathbf{x})$ represent the amount of mineral that can be found at $\mathbf{x}$. Drilling probe holes to investigate the quality and amount of minerals can incur significant costs in terms of machinery, man power, and expenses. Instead, if fewer sites are dedicated as probe sites and, based on the presence of minerals at these locations, deductions can be made regarding the overall presence of minerals over the entire region, then the costs incurred will be much lower by doing so. Thus, if a surrogate function $\hat{f} \approx f$ is used to approximate the true cost function, this procedure can be very economical. In other words, this requires a *prior* over the target space and an optimization routine, which together are referred to as Bayesian Optimization (BO). The pseudo-code for BO is given in Algorithm 1.

Algorithm 1 BayesOpt $(\boldsymbol{\mu}^*, \mathcal{K}^{**}, p(\boldsymbol{\theta}), \mathbf{x}^*, f, \tau, n)$.

1: **Input:**
  - $\boldsymbol{\mu}^*$: prior mean function (can be zero-mean)
  - $\mathcal{K}^{**}$: prior covariance kernel function (e.g., RBF, Matern$\frac{1}{2}$, etc.)
  - $p(\boldsymbol{\theta})$: prior over hyper-parameters
  - τ : acquisition function for sampling
  - n : number of samples to be acquired
  - $\mathbf{x}^*$: test inputs for interpolation
  - f : original function to be modeled

2: **Output:**
  - D_{max} : global maxima and its index for f
  - $\boldsymbol{\mu}_{f|D}$: posterior mean
  - $\mathcal{K}_{f|D}$: posterior covariance

3: $D = []$ ▷ For storing observations
4: $i \leftarrow 0$
5:
6: **do**
7:     $x^+{}_{i+1} \leftarrow \arg\max_x \tau(x, \{x_i\}, \{y_i\}, \theta_{i+1})$ ▷ Next location
8:     $y^+{}_{i+1} \leftarrow f(x^+{}_{i+1})$ ▷ Query objective function
9:     $D \leftarrow D \cup \{x^+{}_{i+1}, y^+{}_{i+1}\}$ ▷ Augment data
10:     $i \leftarrow i + 1$
11:
12: **while** $i < n$
13: $\boldsymbol{\mu}_{f|D}, \mathcal{K}_{f|D} \leftarrow \text{GPR}(D[0, :], D[1, :], \mathcal{K}(\cdot, \cdot), p(\boldsymbol{\theta}), \mathbf{x}^*)$ ▷ Posterior
14: $D_{max} \leftarrow \max(D)$ ▷ Get global maxima and index
15: **return** $\boldsymbol{\mu}_{f|D}, \mathcal{K}_{f|D}, D_{max}$

The details of this algorithm are as follows: the algorithmic inputs include the prior mean and covariance of GP given by $\boldsymbol{\mu}^*, \mathcal{K}^{**}$, respectively. A prior distribution over the kernel parameters $p(\boldsymbol{\theta})$ encodes the prior model of the target space, and τ is a pre-selected information acquisition function sampling from which is cheaper than directly sampling from f; n represents the number of samples that need to be acquired which is also used as a termination criterion. The algorithm begins by selecting the next sample to observe (line 7) and the corresponding observation is acquired by querying the original function f (line 8). The input location and the corresponding observation are then accumulated as D as shown in line 9. Upon termination, the GP posterior is generated (line 13) and the global maxima D_{max} is returned (line 14).

For further details especially regarding the acquisition functions (τ), the readers are referred to [29, Chap. 2] and other theoretical works like [30–34].

6.9 Hands-on experience with GPs

In order to provide the readers with a hands-on experience with GPs and their application to the regression problem, a crash course has been designed and presented in [35]. As a quick summary, the pseudo-code is provided below in Algorithm 2.

The details of this algorithm are as follows: to perform GPR, the algorithmic inputs required are the training inputs ($\mathbf{x}^-$) and corresponding training targets ($\mathbf{y}^-$). Additionally, a covariance of choice needs to be defined via the $\mathcal{K}(\cdot, \cdot)$. An initial guess of the hyper-parameters ($\boldsymbol{\theta}_{init}$) is needed to initialize the kernel(s). Aside from this, a set of test inputs ($\mathbf{x}^*$) is provided for which a posterior can be generated. The posterior

---

Algorithm 2 GPR ($\mathbf{x}^-, \mathbf{y}^-, \mathcal{K}(\cdot, \cdot), \boldsymbol{\theta}_{init}, \mathbf{x}^*$).

1: **Input:**
- $\mathbf{x}^-$: training Inputs (nodes/locations)
- $\mathbf{y}^-$: targets (measurements/observations)
- $\mathcal{K}(\cdot, \cdot)$: covariance kernel function (e.g., RBF, Matern$\frac{1}{2}$, etc.)
- $\boldsymbol{\theta}_{init}$: initial guess about hyper-parameters
- $\mathbf{x}^*$: test inputs for interpolation

2: **Output:**
- $\boldsymbol{\mu}_{f|D}$: posterior mean
- $\mathcal{K}_{f|D}$: posterior covariance
- $-\mathcal{LL}$: negative log-marginal likelihood

3: $[\sigma_s, l_s, \sigma_n] \leftarrow \boldsymbol{\theta}_{init}$
4: $\mathcal{K} = \mathcal{K}(\mathbf{x}^-, \mathbf{x}^-)$ ▷ Compute necessary covariances
5: $\mathcal{K}^* = \mathcal{K}(\mathbf{x}^*, \mathbf{x}^-)$
6: $\mathcal{K}^{**} = \mathcal{K}(\mathbf{x}^*, \mathbf{x}^*)$
7: $\mathcal{K}_\epsilon = \mathcal{K} + \sigma_n^2 \times \mathcal{I}$ ▷ Noisy observations
8: $L = \text{CHOLESKY}(\mathcal{K}_\epsilon)$ ▷ Cholesky decomposition for matrix inversion
9: $m = L^T \backslash (L \backslash \mathbf{y}^-)$ ▷ Solve matrix equation
10: $\mathcal{LL} = 0.5 * (\{\mathbf{y}^-\}^T * m - |\#(\mathcal{K})| \log(2\pi)) - \sum(\text{DIAG}(\log(L)))$ ▷ LML
11: $v = L \backslash \mathcal{K}^*$
12: $\boldsymbol{\mu}_{f|D} = \mathcal{K}^* * m$ ▷ Posterior mean
13: $\mathcal{K}_{f|D} = \mathcal{K}^{**} - v^T v$ ▷ Posterior covariance
14: **return** $\boldsymbol{\mu}_{f|D}, \mathcal{K}_{f|D}, -\mathcal{LL}$

mean is represented by $\mu_{f|D}$ and the corresponding covariance is given by $\mathcal{K}_{f|D}$. Before the regression progresses, it is essential to define the necessary kernels which were previously defined in depth above, and are done so algorithmically via lines 4–7. As mentioned earlier on, inversion of kernel is slow and numerically unstable, Cholesky decomposition is therefore utilized for factorizing the kernel into lower Cholesky factor (line 8). The algorithm then progresses by performing algebraic computations and evaluating the log-marginal likelihood which is done in line 10. This gives rise to the predictions over the test inputs which are obtained in lines 12–13, which are finally returned by the algorithm in line 14.

The key factor to note here is that the entire regression algorithm simply progressed by assuming a fixed set of hyper-parameters (θ_{init}). However, as more data becomes available, it might in fact be useful to tune the hyper-parameters to account for over- and under-fit observations. This involves performing optimization which was done via the type-II maximum likelihood estimation approach. This algorithm is summarized in Algorithm 3 and involves the covariance structure $\mathcal{K}(\cdot, \cdot)$, initial guess of parameters that best define the kernel before observations are gathered (θ_{init}), an upper-bound on maximum number of restarts ($Nres$) that the algorithm should perform to avoid getting caught in locally optimal solutions and lastly, *bounds* which define the upper and lower bounds on each of the hyper-parameters so that they can be appropriately resampled when restarting the search for optimal solution. First, a random set of initial guess of hyper-parameters is initialized (line 7) and this process is done each time a new restart is initialized. This allows for optimization from different starting configuration which allows for exploring different parts of the likelihood surface and evading local optima. Then, the log-likelihood is computed (line 9) and component-wise derivatives are obtained in line 10. These analytical gradients are then fed to an optimizer of choice (line 11) which performs line-searches until a termination condition is reached and returns an optimal set of hyper-parameters. These intermediate log-likelihood and optimal solutions are stored over all restarts (line 12). Upon reaching the termination criterion, the maximum log-likelihood and optimal set of hyper-parameters are extracted as shown in lines 15–16. Finally, the overall results are returned via line 17.

Once, the optimal set of hyper-parameters are obtained, the kernels in lines 4–7 of Algorithm 2 can be reevaluated. This will update their respective entries which will eventually reflect in the updated posterior shown in lines 12–13.

6.10 Pitfalls

Besides the limitations mentioned above, the overall Bayesian inference mechanism itself suffers from a major pitfall, viz., *local optimality*. As the inference method relies on gradient descent methods, the solution obtained may converge to a local optimum.

Algorithm 3 MLE ($\mathcal{K}(\cdot, \cdot), \boldsymbol{\theta}_{init}, Nres, bounds$).

1: **Input:**

- $\mathcal{K}(\cdot, \cdot)$: covariance kernel function (e.g., RBF, Matern$\frac{1}{2}$, etc.)
- $\boldsymbol{\theta}_{init}$: initial guess of the hyper-parameters of $\mathcal{K}(\cdot, \cdot)$
- $Nres$: number of restarts required for optimization
- $bounds$: upper and lower bounds for hyper-parameters in the form $[[UB, LB], [UB, LB], \ldots]$

2: **Output:**

- $\boldsymbol{\theta}$: optimized hyper-parameters

3: $steps = 0$ ▷ Initialize step counter

4: $\boldsymbol{\theta}_{res} = [], \mathbf{ll}_{res} = []$ ▷ For storing results

5:

6: **do**

7:     $\boldsymbol{\theta}_{init} \leftarrow RandomGen(\boldsymbol{\theta}_{init}, bounds)$ ▷ Bounded random initial guesses

8:     $steps += 1$

9:     $\mathcal{LL} = \underbrace{-\frac{n}{2}\log(2\pi) - \frac{n}{2}\log(\sigma_{sig}{}^2)}_{\text{Const.}} - \underbrace{\frac{1}{2}\log|\mathcal{K}|}_{\text{Complexity}} - \underbrace{\frac{1}{2\sigma_{sig}{}^2}\mathbf{y}^T\mathcal{K}^{-1}\mathbf{y}}_{\text{Data fit}}$

10:     $\frac{\partial}{\partial\theta_j}(\mathcal{LL}) = \frac{1}{2}\mathbf{y}^T\mathcal{K}^{-1}\mathbf{y}\frac{\partial\mathcal{K}}{\partial\theta_j}\mathcal{K}^{-1}\mathbf{y} - \frac{1}{2}\text{tr}\left(\frac{\partial\mathcal{K}}{\partial\theta}\right)$ ▷ Component-wise derivatives

11:     $\boldsymbol{\theta}_{interim} \leftarrow Optimize(\frac{\partial\mathcal{LL}}{\partial\theta})$ ▷ Optimization

12:     $\boldsymbol{\theta}_{res} \leftarrow \cup\{\boldsymbol{\theta}_{interim}\}, \mathbf{ll}_{res} \leftarrow \cup\{-\mathcal{LL}\}$ ▷ Accumulate all results

13:

14: **while** $steps < Nres$

15: $bestIdx \leftarrow max(\mathbf{ll}_{res})$ ▷ Index of max log-likelihood

16: $\boldsymbol{\theta} \leftarrow \boldsymbol{\theta}_{res}[bestIdx]$ ▷ Best set of parameters

17: **return** $\boldsymbol{\theta}, -\mathcal{LL}$

This can be easily seen by eye-balling Eq. (6.6) which is a non-convex function [36], optimizing over which will lead to local optima.

This is a well-known pitfall for gradient based methods. In order to overcome such problems, either gradient-free methods could be explored or the optimizer could be restarted from multiple random initial starting conditions. The former has already started to gain traction as presented in works like [37] and [38] investigating theoretical lower-bounds on performance of derivative free optimization (DFO), while, for the latter, it begs the question: *How many restarts would suffice?* This is an extensive research problem in itself and has been explored, for instance, in [39]. Besides this work, there are few other works that have attempted such a feat, which goes without saying how challenging it can be to devise the optimal stopping criterion. However, as remarked in Chap. 5, this will not be investigated further in this book.

6.11 Summary

Given the length of this chapter, the readers would have already realized the level of significance of this chapter. This chapter is one of the foundation chapters for the theme of this book – *Environment Monitoring*. Herein, the intuitive and mathematical notions of the most widely used non-parametric Bayesian method called Gaussian Process (GP) were discussed. Several technical terms pertaining to the GPs along with the applications were described with regards to *classification*, *optimization*, and *regression* setting. A hands-on crash course was also described should the readers decide to do a deep-dive and get to grips with the implementation of this mathematical model. While this crash-course was not meant to be computationally optimal and competitive with regards to the open-source GP libraries, it was purely meant to expose the GPs for what they are and not just being used as a black-box method which takes in training samples and spits out a posterior. Several pitfalls to these methods were also described. To this end, the readers were exposed to both the good and the bad when it comes to GPs and the author(s) believe that by now, the readers would start to see why the GPs are the model of choice for this book and *environmental monitoring* in general. With this, the basic machine learning aspect of this book has been covered, and in what follows the robotic aspects will be discussed. More concretely, this chapter gave a model but the next two chapters discuss how one could gather training samples for GPs without having to manually decide which samples are good for training and which are to be used for testing purposes.

References

[1] C.E. Rasmussen, C.K.I. Williams, Gaussian Processes for Machine Learning (Adaptive Computation and Machine Learning), The MIT Press, 2005.

[2] P. Abrahamsen, A Review of Gaussian Random Fields and Correlation Functions, 1997.

[3] M.L. Stein, Space–time covariance functions, Journal of the American Statistical Association 100 (469) (2005) 310–321.

[4] D. Duvenaud, Automatic Model Construction With Gaussian Processes, Ph.D. thesis, University of Cambridge, 2014.

[5] K. Tiwari, Multi-Robot Resource Constrained Decentralized Active Sensing for Online Environment Monitoring, Ph.D. thesis School of Information Science, Japan Advanced Institute of Science and Technology (JAIST), 2018.

[6] M.G. Genton, Classes of kernels for machine learning: a statistics perspective, Journal of Machine Learning Research 2 (Dec) (2001) 299–312.

[7] R.P. Adams, O. Stegle, Gaussian process product models for nonparametric nonstationarity, in: Proceedings of the 25th International Conference on Machine Learning, ACM, 2008, pp. 1–8.

[8] S. Garg, A. Singh, F. Ramos, Learning non-stationary space-time models for environmental monitoring, in: Proceedings of the AAAI Conference on Artificial Intelligence, vol. 25, 2012, p. 45.

[9] A. Gretton, K. Fukumizu, C.H. Teo, L. Song, B. Schölkopf, A.J. Smola, A kernel statistical test of independence, in: Advances in Neural Information Processing Systems, 2008, pp. 585–592.

[10] B. Fischer, N. Gorbach, S. Bauer, Y. Bian, J.M. Buhmann, Model selection for Gaussian process regression by approximation set coding, arXiv preprint, arXiv:1610.00907, 2016.

[11] A.B. Abdessalem, N. Dervilis, D.J. Wagg, K. Worden, Automatic kernel selection for Gaussian processes regression with approximate Bayesian computation and sequential Monte Carlo, Frontiers in Built Environment 3 (2017) 52.

[12] A. Blumer, A. Ehrenfeucht, D. Haussler, M.K. Warmuth, Occam's razor, Information Processing Letters 24 (6) (1987) 377–380.

[13] J. Chen, K.H. Low, Y. Yao, P. Jaillet, Gaussian process decentralized data fusion and active sensing for spatiotemporal traffic modeling and prediction in mobility-on-demand systems, IEEE Transactions on Automation Science and Engineering 12 (3) (2015) 901–921.

[14] P. Boyle, M. Frean, Dependent Gaussian processes, in: Advances in Neural Information Processing Systems, 2005, pp. 217–224.

[15] M.A. Osborne, S.J. Roberts, A. Rogers, S.D. Ramchurn, N.R. Jennings, Towards real-time information processing of sensor network data using computationally efficient multi-output Gaussian processes, in: Information Processing in Sensor Networks, 2008. IPSN'08. International Conference on, IEEE, 2008, pp. 109–120.

[16] P. Moreno-Muñoz, A. Artés-Rodríguez, M.A. Álvarez, Heterogeneous multi-output Gaussian process prediction, arXiv preprint, arXiv:1805.07633, 2018.

[17] I. Bilionis, N. Zabaras, Multi-output local Gaussian process regression: applications to uncertainty quantification, Journal of Computational Physics 231 (17) (2012) 5718–5746.

[18] H. Liu, Y.S. Ong, X. Shen, J. Cai, When Gaussian process meets big data: a review of scalable GPs, arXiv:1807.01065, 2018.

[19] K. Chalupka, C.K. Williams, I. Murray, A framework for evaluating approximation methods for Gaussian process regression, Journal of Machine Learning Research 14 (Feb) (2013) 333–350.

[20] R. Herbrich, N.D. Lawrence, M. Seeger, Fast sparse Gaussian process methods: the informative vector machine, in: Advances in Neural Information Processing Systems, 2003, pp. 625–632.

[21] S. Oh, Y. Xu, J. Choi, Explorative navigation of mobile sensor networks using sparse Gaussian processes, in: Decision and Control (CDC), 2010 49th IEEE Conference on, IEEE, 2010, pp. 3851–3856.

[22] David M. Blei, Alp Kucukelbir, Jon D. McAuliffe, Variational inference: a review for statisticians, Journal of the American Statistical Association 112 (518) (2017) 859–877.

[23] James M. Joyce, Kullback-leibler divergence, International Encyclopedia of Statistical Science (2011) 720–722.

[24] Michalis Titsias, Variational learning of inducing variables in sparse Gaussian processes, in: Artificial Intelligence and Statistics, 2009, pp. 567–574.

[25] Ching-An Cheng, Byron Boots, Variational inference for gaussian process models with linear complexity, in: Advances in Neural Information Processing Systems, 2017, pp. 5184–5194.

[26] D. Bui Thang, V. Nguyen Cuong, Siddharth Swaroop, Richard E. Turner, Partitioned Variational Inference: a unified framework encompassing federated and continual learning, arXiv:1811.11206, 2018.

[27] Robert B. Gramacy, Daniel W. Apley, Local Gaussian process approximation for large computer experiments, Journal of Computational and Graphical Statistics 24 (2) (2015) 561–578.

[28] R. Marchant, F. Ramos, Bayesian optimisation for intelligent environmental monitoring, in: Intelligent Robots and Systems (IROS), 2012 IEEE/RSJ International Conference on, IEEE, 2012, pp. 2242–2249.

[29] R. Marchant Matus, Bayesian Optimisation for Planning in Dynamic Environments, 2015.

[30] E. Brochu, V.M. Cora, N. De Freitas, A tutorial on Bayesian optimization of expensive cost functions, with application to active user modeling and hierarchical reinforcement learning, arXiv preprint, arXiv:1012.2599, 2010.

[31] E. Brochu, M.W. Hoffman, N. de Freitas, Portfolio allocation for Bayesian optimization, arXiv preprint, arXiv:1009.5419, 2010.

[32] D.J. Lizotte, Practical Bayesian Optimization, University of Alberta, 2008.

[33] M.A. Osborne, Bayesian Gaussian Processes for Sequential Prediction, Optimisation and Quadrature, Ph.D. thesis, Oxford University, UK, 2010.

[34] M. Schonlau, Computer Experiments and Global Optimization, 1997.

[35] K. Tiwari, Hands-on experience with Gaussian processes (GPs): implementing GPs in Python-I, arXiv preprint, arXiv:1809.01913, 2018.

[36] A. Melkumyan, F. Ramos, A sparse covariance function for exact Gaussian process inference in large datasets, in: IJCAI, vol. 9, 2009, pp. 1936–1942.

[37] P. Koch, O. Golovidov, S. Gardner, B. Wujek, J. Griffin, Y. Xu, Autotune: a derivative-free optimization framework for hyperparameter tuning, in: Proceedings of the 24th ACM SIGKDD International Conference on Knowledge Discovery & Data Mining, ACM, 2018, pp. 443–452.

[38] K.G. Jamieson, R. Nowak, B. Recht, Query complexity of derivative-free optimization, in: Advances in Neural Information Processing Systems, 2012, pp. 2672–2680.

[39] T. Dick, E. Wong, C. Dann, How Many Random Restarts Are Enough?, Google Scholar, 2014.

Coverage Path Planning (CPP)
Maximal area coverage

7

One way to acquire knowledge is to know everything about everything you don't know. But there is only so much your brain can take.

Dr. Kshitij Tiwari

Contents

Highlights

- Operational definition of Coverage Path Planning (CPP)
- Related applications from combinatorics
- Related applications from robotics
- Related applications from wireless sensor networks
- Challenges

Coverage Path Planning (CPP) is a problem that involves exhaustively covering the area of interest. Such a path planner falls under the *exhaustive area coverage* paradigm where visiting everywhere is of utmost importance. Such methods are useful for several applications, for instance, consider the task of mowing a lawn or vacuuming the room. Given the laborious and mundane nature of the task, people are increasingly switching to utilizing robots as shown in [1]. The operational definition of the CPP is given in Definition 7.1.

Multi-Robot Exploration for Environmental Monitoring. https://doi.org/10.1016/B978-0-12-817607-8.00020-4

Definition 7.1 (Coverage Path Planning (CPP)). Under CPP, the robot/agent is required to visit **all points** for **complete** coverage of the target environment. While doing so, it must follow these requirements [2]:

- The paths must be non-overlapping.
- The paths must be non-repetitive.
- If present, all obstacles must be avoided.
- Simple trajectories must be used like straight lines, circles, etc.
- "Optimal" path under the mission constraints is desired.

In what follows, the CPP is described from various viewpoints: combinatorics, robotics, and wireless sensing. Research in each of these domains has progressed at different rates but this problem persists in several different research settings.

7.1 Coverage path planning in combinatorics

The research problems/applications below illustrate the need for studying and investigating CPP. Albeit these applications from the combinatorics domain also hold a stark resemblance to the CPP problem, there are slight differences which set them apart. Such distinctions are clearly highlighted as the discussion progresses.

7.1.1 Covering Salesman Problem (CSP)

Introduced by Current and Shilling in 1989, the Covering Salesman Problem (CSP) [3] is a slight variation to the well-researched traveling salesman problem (TSP) [4]. In TSP, the salesman needs to travel through each city to sell a product (shown in Fig. 7.1(A)), while in CSP the salesman needs to visit a neighborhood of each city (shown in Fig. 7.1(B)). This, in a CPP setting, would mean that the salesman needs to visit all neighborhoods of all cities (shown in Fig. 7.2). This significantly increases the complexity of the path planning problem as the TSP is already known to be NP-complete [5].

7.1.2 Piano mover's problem

The piano mover's problem as the name suggests is the problem faced when transporting a piano from the shop to the client's home (shown in Fig. 7.3). In this setting, it is assumed that the map of the suburb within which the piano has to be transported is known *a priori*. The challenge then is to find a collision-free path from the shop to the client's home [6]. This problem again has been shown to be NP-hard [7, Chap. 4] and differs from CPP in the sense that the movers do not need to visit every point in the neighborhood to reach the client's home.

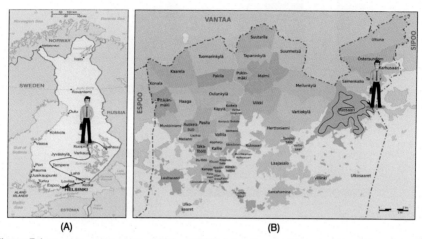

Figure 7.1 Illustration of the related TSP and CSP setting from combinatorics. Fig. 7.1(A) shows the regional map of Finland, while Fig. 7.1(B) shows the city map of Southern Finland. Both figures show the corresponding trajectories of the salesmen under the TSP and CSP setup: (A) Traveling Salesman Problem (TSP), (B) Covering Salesman Problem (CSP).

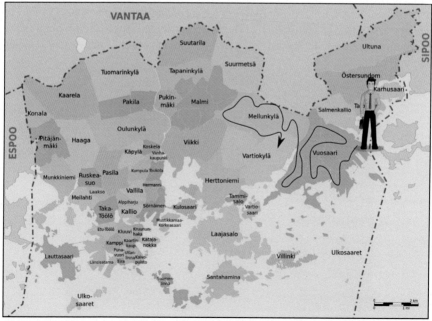

Figure 7.2 Illustration of the CPP salesman scenario. A salesman is traversing each locality of each city to sell his product.

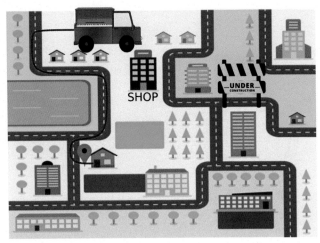

Figure 7.3 A typical piano mover problem. A piano needs to be transported from the shop to the customers home given the map of the town. Some roads may be under constructions while others may have traffic jams. The best route needs to be calculated under such constraints.

7.1.3   Art gallery problem

Consider the scenario where an art gallery hosts an art exhibition with expensive art on display. For security, the art gallery must consult a security firm to hire a set of guards to monitor the gallery at all times. This begs the question: *How many guards are necessary, and how many are sufficient to patrol the paintings and works of art in a gallery with n walls?* [8, Chap. 22]. The key constraint here is that each point in the gallery must be visible to at least one guard so that no corner of the gallery remains unguarded. (See Fig. 7.4.)

Figure 7.4 Illustrating an art gallery with expensive sculptures. The task then is to deploy optimal number of guards to patrol every possible nook-and-cranny of this gallery. Photo based on Shvets Anna from Pexels [9].

7.1.4 Watchman route problem

The watchman route problem [10] refers to the task of finding the shortest route (closed curve) in a given simple polygon such that every point of it is visible from at least one point of the watchman's route. (See Fig. 7.5.)

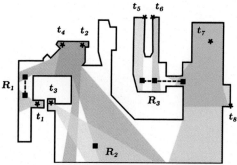

Figure 7.5 Illustration of the watchman routing problem as previously presented in [11]. Consider a set of target points (denoted by $\star$) in a polygonal environment as shown. The objective of the watchmen denoted by $R_1, \dots, R_3$ is to find paths (denoted by $- - -$) such that each target ($\star$) is seen from at least one viewpoint ($\blacksquare$). In this setup, watchman R_1 can observe t_1 and t_2, R_2 can monitor t_3 and t_4 while R_3 can see t_5–t_8.

7.1.5 Orienteering Problem (OP) for vehicle routing

In this setting, all nodes including the start, intermediate, and end nodes are specified *a priori* [12,13]. The intermediate nodes have associated scores which could either be *stationary* or *time-varying* [14]. In [15], another variation of the OP was considered wherein the scores of intermediate nodes were correlated with those of the peers. The challenge here again is, given the limited time available to reach the goal node from the start, an optimal path needs to be generated so as to maximize the total path score. The potential application of such kinds of model is widely found in the Vehicle Routing Problem (VRP) where a fleet of vehicles, all starting at the same location, must be optimally routed to supply the demands of the customers subject to vehicle capacity constraints [12].

7.2 Coverage path planning in robotics

The previous section summarized some of the well-known strategies from combinatorics that closely relate to CPP. This section discusses the related exploration strategies for robotics applications wherein an explicit/entire map of the region of operation need not necessarily be known.

7.2.1 Lawn mower strategy

The lawn mower problem as illustrated in Fig. 7.6 represents the problem of optimally mowing a lawn. Optimality, in loose terms, can be defined for this setting as straight line paths with no overlaps. This problem is already proven to be NP-hard [16] and can only account for static obstacles, if any.

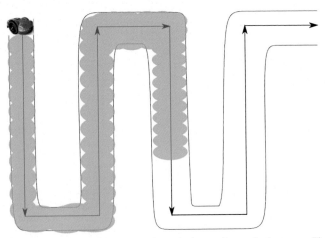

Figure 7.6 Typical lawn mower pattern to mow the entire lawn using a robotic mower. The shaded green area indicates the area that has already been mowed.

7.2.2 Frontier based exploration

This strategy primary relies on the concept of *frontiers* which are defined in Definition 7.2 [17]. In this strategy, as the robot knows almost everything within the sensing range, it tries to reach the frontier so that it extends its map into new territory (as shown in Fig. 7.7). Repeating this behavior for a certain amount of time will eventually lead to the exploration of the entire environment.

Definition 7.2 (Frontier). Given that each sensor has a sensing range within which it can acquire the observations, all the region encompassed by this sensing range is largely known (with some uncertainty owing to sensing noise). Thus, the distinguishing boundary (semantic and not physical) between the open space and the unexplored area is called a frontier.

7.2.3 Voronoi tessellation based area coverage

In [18], a persistent environment monitoring approach was presented for a team of robots whilst accounting for sensing and actuation variations. This approach was based on weighted Voronoi partitioning where the weights represent adaptations in partitions

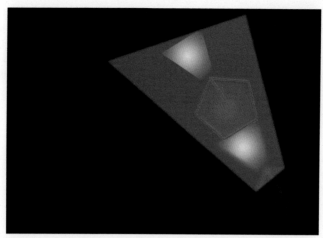

Figure 7.7 Illustration of a frontier-based exploration of a toy robot. Consider a robot equipped with a laser range finder. The area to be explored is largely unknown as shown by the black background. Let the robot start the exploration from the bottom right corner with the laser sensing range shown by shaded white triangle. All the region within this sensing range is known and is open (shown by white shading, free of obstacles). Thus, the robot selects a point along this frontier, moves to that point, and expands its horizon. After this iteration, the area shown by translucent white triangle is now known, while the rest of black region remains unexplored. This procedure can be repeated over and over again, and eventually the whole region will turn from black to white, i.e., unexplored to explored.

of the Voronoi tessellations such that, despite sensing and actuation variations, the entire arena can be monitored at all times. This is illustrated in Fig. 7.8.

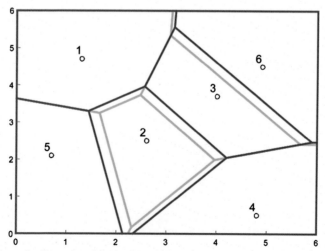

Figure 7.8 A regular Voronoi partition is shown in blue, and the weighted partition is shown in green. Robot 2 has a lower weight, which gives it a smaller cell. Similarly, robot 6 has a higher relative weight, and therefore has a larger cell. Image taken from [18].

As an alternate application to Voronoi partitioning for persistent environment monitoring, in [19] a sporadic asynchronous one-to-base communication architecture was proposed to use a cloud-based infrastructure for controlling a multi-robot team. Coverage assignments and surveillance parameters are managed on the cloud and transmitted to mobile agents via unplanned and asynchronous exchanges. This setting is shown in Fig. 7.9.

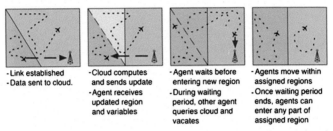

| - Link established
- Data sent to cloud. | - Cloud computes
and sends update
- Agent receives
updated region
and variables | - Agent waits before
entering new region
- During waiting
period, other agent
queries cloud and
vacates | - Agents move within
assigned regions
- Once waiting period
ends, agents can
enter any part of
assigned region |

Figure 7.9 Illustration of the proposed strategy. The partitioning component (executed on the cloud) manages coverage regions and introduces logic to prevent collisions, while the trajectory planning component (executed on-board each agent) governs agent motion. Image taken from [19].

7.3 Coverage path planning in Wireless Sensor Networks (WSNs)

Each wireless sensor has a certain region that it can sense. This region is often called the *sensing range*. A collection of such wireless sensors is then called a wireless sensor network if the information being sensed is being shared amongst them or processed collectively at a base station. For instance, consider Fig. 7.10 wherein a collection of satellites are monitoring the Earth from outer space. In today's era of space race, this is a very common application of WSNs that most people are quite familiar with. This is how Google is able to show maps of almost all regions on Google maps that people seamlessly then use for navigation tasks. The path planning task here is to be able to

Figure 7.10 A collection of satellites working as a wireless sensor network to cover the globe from outer space.

position and optimize the trajectories of the satellites in the orbit such that the global coverage is persistently provided.

7.4 Challenges

Whilst the problem categories discussed here are primarily motivated by coverage, persistent monitoring, and surveillance applications, there is yet another category of problems which does not consider wide area coverage for full-region surveillance. Such problems are what will be referred to as *Informative Path planning (IPP)* problems and will be discussed in the next chapter. There is a stark contrast in the objectives of CPP and IPP, and it is imperative that this difference is clarified.

References

[1] M. Bosse, N. Nourani-Vatani, J. Roberts, Coverage algorithms for an under-actuated car-like vehicle in an uncertain environment, in: Robotics and Automation, 2007 IEEE International Conference on, IEEE, 2007, pp. 698–703.

[2] E. Galceran, M. Carreras, A survey on coverage path planning for robotics, Robotics and Autonomous Systems 61 (12) (2013) 1258–1276.

[3] J.R. Current, D.A. Schilling, The covering salesman problem, Transportation Science 23 (3) (1989) 208–213.

[4] A. Orman, H.P. Williams, A survey of different integer programming formulations of the travelling salesman problem, in: Optimisation, Econometric and Financial Analysis, vol. 9, 2006, pp. 93–108.

[5] G.J. Woeginger, Exact algorithms for NP-hard problems: a survey, in: Combinatorial Optimization—Eureka, You Shrink!, Springer, 2003, pp. 185–207.

[6] G.E. Jan, T.Y. Juang, J.D. Huang, C.M. Su, C.Y. Cheng, A fast path planning algorithm for piano mover's problem on raster, in: Advanced Intelligent Mechatronics. Proceedings, 2005 IEEE/ASME International Conference on, IEEE, 2005, pp. 522–527.

[7] S.M. LaValle, Planning Algorithms, Cambridge University Press, 2006.

[8] J. Urrutia, Art gallery and illumination problems, in: Handbook of Computational Geometry, Elsevier, 2000, pp. 973–1027.

[9] https://www.pexels.com/photo/statue-room-2574476/.

[10] X. Tan, B. Jiang, An improved algorithm for computing a shortest watchman route for lines, Information Processing Letters 131 (2018) 51–54.

[11] P. Tokekar, V. Kumar, Visibility-based persistent monitoring with robot teams, in: Intelligent Robots and Systems (IROS), 2015 IEEE/RSJ International Conference on, IEEE, 2015, pp. 3387–3394.

[12] B.L. Golden, S. Raghavan, E.A. Wasil, The Vehicle Routing Problem: Latest Advances and New Challenges, vol. 43, Springer Science & Business Media, 2008.

[13] P. Vansteenwegen, W. Souffriau, D. Van Oudheusden, The orienteering problem: a survey, European Journal of Operational Research 209 (1) (2011) 1–10.

[14] Z. Ma, K. Yin, L. Liu, G.S. Sukhatme, A spatio-temporal representation for the orienteering problem with time-varying profits, in: Intelligent Robots and Systems (IROS), 2017 IEEE/RSJ International Conference on, IEEE, 2017, pp. 6785–6792.

[15] J. Yu, M. Schwager, D. Rus, Correlated orienteering problem and its application to persistent monitoring tasks, IEEE Transactions on Robotics 32 (5) (2016) 1106–1118.

[16] E.M. Arkin, S.P. Fekete, J.S. Mitchell, Approximation algorithms for lawn mowing and milling, Computational Geometry 17 (1–2) (2000) 25–50.

[17] B. Yamauchi, A frontier-based approach for autonomous exploration, in: Computational Intelligence in Robotics and Automation, 1997. CIRA'97. Proceedings. 1997 IEEE International Symposium on, IEEE, 1997, pp. 146–151.

[18] A. Pierson, L.C. Figueiredo, L.C. Pimenta, M. Schwager, Adapting to sensing and actuation variations in multi-robot coverage, The International Journal of Robotics Research 36 (3) (2017) 337–354.

[19] J.R. Peters, S. Wang, A. Surana, F. Bullo, Cloud-supported coverage control for persistent surveillance missions, Journal of Dynamic Systems, Measurement, and Control (2017).

Informative Path Planning (IPP)
Informative area coverage

The conservation of natural resources is the fundamental problem. Unless we solve that problem it will avail us little to solve all others.

Theodore Roosevelt, 1907

Contents

Highlights

- Sniffing out Information
- Pure Explorative Behavior
- Pure Exploitative Behavior
- Exploration–Exploitation trade-off
- Heeding the Robot Constraints

Chap. 6 introduced the model (GP) that is widely used to explain the complex dynamics of the environmental phenomenon. Just like any other machine learning framework, this model too needs training samples which can either be acquired by an artificial dataset or via a robot during a real mission. While the former is usually handled by a human supervisor, Chap. 7 and this chapter present contrasting information acquisition strategies focused primarily on autonomous robots.

Multi-Robot Exploration for Environmental Monitoring. https://doi.org/10.1016/B978-0-12-817607-8.00021-6

This chapter serves to complement the previously mentioned GP model by endowing the robot(s) with the capability to autonomously plan paths through waypoints (informative locations) while paying heed to the resource constraints. As was exclaimed by President Theodore Roosevelt, ignoring the conservation of resources belittles the significance of solutions proposed for all other problems. Thus, here the author(s) address the problem of resource constrained informative path planning (alternatively, resource constrained active sensing) by proposing a novel information acquisition function that can elegantly trade-off the quality of information gathered to the residual resources without significantly compromising on either.

As done previously, this discussion progresses by considering a real-life example. Consider the example shown in Fig. 8.1. In this setting, the K9 is expected to find the incendiary device in the shortest possible time (this could be considered as a resource constraint) given the risk associated with unexploded incendiary device. In the scene, for the sake of argument, there are only two wooden boxes. One box consists of an incendiary device and the other consists of fruits. Both boxes give out odors (odor represents information here), however, this K9 is especially trained to distinguish the smell of explosives (useful information) from that of edible items (similar to noise). Thus, completely unaware about the environment *a priori*, the K9 follows the odor to try to get closer to the box which gives out a strong odor of the incendiary device. More formally, the K9 follows the gradient of information to reduce the search space. For the scope of this book, such a path planning strategy will be referred to as Informative Path Planning (IPP) or active sensing.

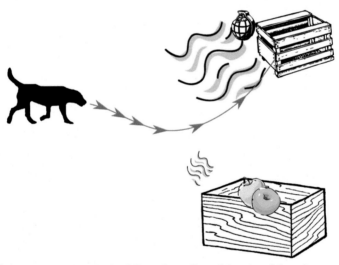

Figure 8.1 Bomb sniffing K9. The dog follows the gradient of the odor of the ordnance as opposed to walking around all over the place which would take considerably longer time.

Over the past decade, a lot of researchers have started to investigate this behavior by focusing on applications utilizing autonomous mobile robots to assist with large-scale environment monitoring [1–4]. The task at hand is challenging, owing to complex dynamics, data-yearning models, and the resource constrained robots which often tend to conflict each other. This chapter introduces the resource constraint variants of the path planning approaches via waypoints.

8.1 Planning over waypoints

Waypoint (cf. Definition 8.1) based path planning for environmental monitoring applications works under the assumption that the time required for a robot to finish its mission is comparatively smaller than the time required by the environment dynamics to evolve. This can also be interpreted in terms of the K9 example from Fig. 8.1 as a case when the K9 starts to follow the odor and suddenly the boxes are moved around or a wind gust disperses the odors. In that case, the K9 would have to stay in one place until it re-acquires the odor and, if these changes are too frequent, it would rather not move at all.

For the robotics case, the reason why such an assumption is critical is that, if the field dynamics were to update faster than the robot motion, then the robot would also choose to stay in its current location as it would be presented with novel information across time without having to move. In light of this assumption, for any particular time-slice t_i, the spatio-temporal waypoint selection for a single robot case could be represented like in Fig. 8.2. In this setting, the set of locations chosen to be observed for one time-slice become independent of those chosen for any other. Similarly, interpolated observations at one time-slice are also independent of those for any other.

Definition 8.1 (Waypoint). A waypoint is an intermediate goal position that must be reached in order to gather observation for updating the GP model. Observations can only be gathered from waypoints and the current robot positions. Of all the potential candidates that may be chosen as the waypoint, only those that return the maximal reward are selected where the reward is evaluated in terms of the amount of information accrued.

8.1.1 Selection of waypoints

Having defined the intuition behind the waypoint based path planning, the next obvious aspect to consider is the selection of waypoints. They can either be pre-meditated, for instance, decided by human supervisor(s), or can be selected on the fly or/and autonomously. As this book focuses on the enhancing the autonomy suit of the mobile robot teams for exploring the environmental phenomena, the latter method is of greater interest and will be discussed further.

In the spirit of using the information to guide the selection of waypoints, it is imperative to clarify the notion of *active sensors*. Active sensors (cf. Definition 8.2) are known to transmit energy and rely on the ray-tracing mechanism to decipher the occupancy of the environment. Thus, an autonomous waypoint selection, or, in other words, *active sensing* (cf. Definition 8.3) would encompass evaluating the informativeness of the locations and then selecting the most informative candidate location.

This process can then be repeated until a termination condition is met and allows for actively and iteratively selecting waypoints while exploring.

Definition 8.2 (Active sensor). As per the remote sensing literature, an *active sensor* is defined as a sensor which *actively* transmits energy in the form of light or electrons to be bounced off by a target. This is the operational principle of sensors like sonars, LIDARs, laser range finders, GPS, X-rays, etc. In this setting, the rays that bounce off the target provide information about the occupied space (occupied by the target).

Definition 8.3 (Active sensing). The process of *actively* guiding the sensor to perceive information about variables of interest in a purposive and information-seeking fashion.

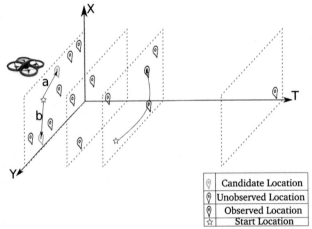

⊚	Candidate Location
⊚	Unobserved Location
⊚	Observed Location
☆	Start Location

Figure 8.2 Planning over waypoints. Illustrating waypoint selection and interpolation mechanism and generalizing over multiple time-steps for a single robot case. The objective of this waypoint-based path planning is to do interpolation, i.e., making accurate predictions at locations marked with red pins. The start location for a particular time-slice is represented by a blue star while all the visited locations are shown with blue pins and the *next-best-location*, i.e. green pins (or choice between candidates a and b) are iteratively chosen. Under the assumptions set above, interpolation across one time-slice t_i is independent of that across $t_j; i \neq j$.

As previously discussed in Chap. 2, information can be broadly categorized into observations presenting *high measurements* or *high variance*. From hereon out, the author(s) focus only on *high variance* based active information acquisition functions which are detailed below.

8.1.2 Notational convention for inputs and targets

The following notational convention will be used: Let *dom* represent the domain of the target phenomenon within which the robot is confined to gather observations. Then, let $U \subset dom$ be the set of unobserved (inputs) locations and let $O \subset dom$ be the set of observed (inputs) locations. Also, assume $\mathbf{x}^*$ represents the locations in the set $U \subset dom$ comprising the set of unobserved locations over which predictions need to be made; $\mathbf{x}^-$ represents the list of visited locations in $O \subset dom$; $\mathbf{x}^+$ represents the candidate locations which can be observed next, while $\hat{\mathbf{x}} \in \mathbf{x}^+$ represents the optimal *next-best-location*. Similarly, $\mathbf{y}^*$, $\mathbf{y}^-$, and $\mathbf{y}^+$ represent the measurements at such locations, respectively.

In what follows, a variety of path planning strategies will be discussed below which range from fully explorative to fully exploitative in nature. Additionally, the novel resource constrained variant is also described at length.

8.1.3   Entropy maximization (full-DAS)

This acquisition function drives the robot to follow the gradient of information to sample observations from areas with high uncertainty or the least knowledge. Mathematically, this is defined by harnessing the entropy ($\mathbb{H}$) of GP as

$$\hat{\mathbf{x}}_F = \arg\max_{\mathbf{x}^+} \left\{ \mathbb{H}_{\mathbf{y}^+|\mathbf{y}^-,\theta} \right\}. \tag{8.1}$$

From hereon, Eq. (8.1) will be referred to as *full-DAS* wherein *DAS* stands for Decentralized Active Sensing. This objective function is shown for a single robot but can be simply scaled to a team of disconnected (cf. Definition 8.5) and decentralized (cf. Definition 8.4) robots which can be used to obtain the purely explorative behavior. However, one major drawback of such an objective function is that the robot tends to visit far off locations thereby incurring prohibitively high travel costs [5].

Definition 8.4 (Decentralized). Within the scope of this book, the term *decentralized* refers to the notion of each agent processing its own data locally (on-board). For an intuitive explanation and further details, the readers are referred to Chap. 12.

Definition 8.5 (Disconnected). For the scope of this work, the author(s) define disconnected as a robot peer-to-peer communication strategy whereby all communication overhead is avoided. In other words, the peers communicate neither with each other nor with the base. This is usually the case for harsh operational environments like sub-surface explorations. Additional details are available in Chap. 12.

8.1.4   Nearest Neighbor (NN)

A naïve solution to remedy the extreme travel costs incurred by *full-DAS* could be to strictly focus on thrifty utilization of resources. In other words, this means visiting the nearest neighbors, which leads to minimal travel costs and renders a fully exploitative nature. But, from the Tobler's first law [6], it is evident that this leads to acquisition of highly correlated observations. Simply put, high correlation translates to redundant or un-informative observations thereby incurring high resource wastage and poor model quality:

$$\hat{\mathbf{x}}_N = \arg\min_{\mathbf{x}^+} \left\{ \ln\|\mathbf{x}^- - \mathbf{x}^+\| \right\}. \tag{8.2}$$

8.1.5   Resource utilization efficacy amelioration while maximizing information gain

In order to avoid incurring extreme travel cost and also refrain from gathering highly correlated observations, this work proposes a novel acquisition function. The new acquisition function trades off the quality of observations (high entropy) to resource optimization (low travel cost) using a bi-objective optimization framework. In [7] the author(s) proposed a novel bi-objective optimization framework which is referred to as *Resource Constrained Decentralized Active Sensing (RC-DAS)*. This approach can effectively trade off model quality to travel distance without compromising the model performance significantly, and furthermore bounds the travel costs. Essentially this problem is difficult to solve owing to the conflicting nature of the objectives: GP models are highly data-driven models, i.e., the larger the number of training samples, the better the predictive performance. But in order to acquire large numbers of "informative" training samples, the robot would incur excessive travel costs. Thus, the challenge is to optimize over both objectives simultaneously without significantly compromising either one. This can be done elegantly using the proposed objective function:

$$\hat{\mathbf{x}}_R = \arg\max_{\mathbf{x}^+} \left\{ \alpha \underbrace{\mathbb{H}_{\mathbf{y}^+|\mathbf{y},\theta}}_{Entropy} - (1-\alpha)\ln\underbrace{\|\mathbf{x}^- - \mathbf{x}^+\|}_{Distance} \right\}. \tag{8.3}$$

In Eq. (8.3), the model performance ($\mathbb{H}_{\mathbf{y}^+|\mathbf{y},\theta}$) and travel cost ($\ln\|\mathbf{x}^- - \mathbf{x}^+\|$) are amalgamated[1] into one bi-objective optimization routine using the weight α. The weight α handles the exploration–exploitation trade-off. The approach is scalable to a multi-robot setting as is summarized in Algorithm 4.

In Algorithm 4, when the robot is at a certain location $x^-{}_m$ and obtains a sensor measurement y^-, the observation (line 9) and input location (line 10) are stored for inference (line 14). Then, the most uncertain neighbors surrounding the current robot location are evaluated, which are within accessible limits of the robot (line 16). Following suit, the posterior prediction over these locations as shown in line 18 is deduced. To evaluate the *most informative next-best-location*, the proposed cost function from Eq. (8.3) is evaluated to optimize the travel distance and simultaneously reduce the prediction uncertainty (line 20). Upon jointly maximizing over this cost function, the feasible *next-best-location* as shown in line 22 can be obtained which is then set as the current goal position to be attained by the mobile robot. In line 28, the remaining budget B is updated by subtracting the *Sensing cost* (S) and *Travel cost* (T) incurred as a result of motion to the new location $\hat{x}_R$. For the purpose of evaluations, the *Sensing cost* (S) and *Travel cost* (T) have been defined by the author in Definitions 8.6 and 8.7, respectively.

---

[1] The readers are cautioned here that the two objective functions are in different domains. In order to amalgamate distance with the entropy, the resultants are first normalized into a dimensionless quantity and then fused together as a weighted combination shown here.

Algorithm 4 RC-DAS (D,B).

1: **Input:**

- $m \in \{1, \dots, M\}$ ▷ Number of robots
- $\{\mathbf{x}^-\}_m \leftarrow \mathbf{x}_m^{-[1]}$ ▷ Initialize with start location
- $\{\mathbf{y}^-\}_m \leftarrow NULL$ ▷ Initialize with null
- $\{O\}_m \leftarrow NULL$

2: **Output:** *Next-best-location, $\hat{x}_R$*

3: **for** agent $m = 1, \dots, M$ **do**

4:      **while** *Budget B >0* **do**

5:

6:      /***SENSE***/

7:

8:      $y^- \leftarrow Sense(x_m^-)$ ▷ Obtain measurement

9:      $\mathbf{y}^-{}_m \leftarrow [\mathbf{y}^-{}_m, ; y]$ ▷ Store observation

10:      $O_m \leftarrow [O_m ; x^-{}_m]$ ▷ Store location

11:

12:      /***PLAN***/

13:

14:      $\hat{\boldsymbol{\theta}}_{opt} \leftarrow MLE(\mathbf{y}^-{}_m, O_m)$ ▷ Obtain hyperparameters

15:      ▷ Deduce most uncertain locations

16:      $\mathbf{x}^+{}_m \leftarrow CalcUncertainNeighbors(D, x_m)$

17:      ▷ Compute predicted measurements

18:      $\mu^+{}_m, \Sigma^+{}_m \leftarrow CompPosterior(\mathbf{z}_m, O_m, \mathbf{x}^+{}_m, \hat{\boldsymbol{\theta}}_{opt})$

19:      ▷ RC-DAS objective function

20:      $Obj \triangleq \left(\alpha\, \mathbb{H}_{\mathbf{y}^+ | \mathbf{y}, \boldsymbol{\theta}} - (1 - \alpha)\ln(D_{Hav}(\mathbf{x}^- - \mathbf{x}^+)) \right)$

21:      ▷ Optimal *Next Best Location*

22:      $\hat{x}_R \leftarrow \arg\max_{\mathbf{x}^+}(Obj)$

23:

24:      /***ACT***/

25:

26:      ▷ Pass target location to robot controller

27:      $x_m^- \leftarrow MoveToNextBestLoc(\hat{x}_R)$

28:      $B \leftarrow B - (S + T)$ ▷ Update remaining budget

29:      **end while**

30: **end for**

Definition 8.6 (Sensing cost). The cost incurred by the robot to gather a measurement. This involves the cost of operating the sensor, processing the measurements, heat-losses from sensors (if any), etc. This cost is usually static and, for this book, was defined as a function of commute from the current location x^- to the candidate location x^+. Mathematically, this becomes

$$C_S(x) = \arg\min_{\forall x^+} ||x^- - x^+||. \tag{8.4}$$

The intuition behind such a choice of sensing cost was that, irrespective of how good the sensor models and energy dissipation models are, the robot will always consume energy (E), $E \geq \arg\min_{\forall x^+} ||x^- - x^+|| + \delta$, where the term δ accounts for all the other aforementioned ancillary losses incurred while operating the sensor.

Definition 8.7 (Travel cost). The cost incurred by the robot to move from its current location to the *next-best-location* for gathering the measurement. This cost encompasses the motor losses, aerial drag losses, losses owing to unforeseen environmental conditions, battery energy consumed to move from one waypoint to another, mechanical losses, etc. This cost is highly dynamic, and accurately modeling it is challenging. For the scope of this work, the following definition of travel cost would suffice:

$$C_T(x, x') \triangleq ||x^- - x^+||. \tag{8.5}$$

8.1.6 Comparative analysis of information acquisition functions

The weight factor α used in *RC-DAS* can be adapted to change its behavior. When $\alpha \to 0$, the path planner behaves like *nearest neighbor (NN)* and when $\alpha \to 1$ it behaves like *full-DAS*. A comparative analysis of all three acquisition functions is visually aided by Fig. 8.3. Besides, the computational complexities of *full-DAS/RC-DAS/NN* are all given by $\mathcal{O}(MN)$ for a team of M robots evaluated over N nodes spread across the domain *dom*.

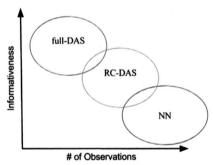

Figure 8.3 A comparison of acquisition functions. Comparing the information acquisition function on a spectrum of highly informative to highly resource constrained options.

Full-DAS is highly explorative in nature; therefore, it gathers a handful of highly "informative" training samples, whilst *nearest neighbor* is highly exploitative forcing the robot to gather highly correlated (un-informative) observations. *RC-DAS* is elegantly positioned between the former two path planning approaches delicately handling the quantity and quality of observations being accrued. As was discussed earlier, GPs are highly data-driven models. So, neither scanty numbers of high quality observations nor excessive amounts of redundant information would be of any significance.

8.2 Homing

Despite considering the resource constraints in *RC-DAS* acquisition function, there remains one severe lacuna yet, i.e., the robot tends to use its resources conservatively but this does not guarantee the robot return to the base station (home) at the end of its mission. This is referred to as "homing" and if not dealt with, it could simply lead to the robot getting immobilized amidst the mission. To this end, a dynamic weight selection mechanism is proposed which not only updates the weights of the objective functions of *RC-DAS* as the resources are being used, but also explicitly takes the homing issue into consideration.

Definition 8.8 (Homing). The ability of a robot to successfully execute a trajectory such that it is always guaranteed to return to the base station (hardware failures not accounted for) is referred to as homing. Homing is essential for two reasons:

- **Reduce Phase.** For the next phase of the architecture (details in Chap. 13) wherein all models generated by various robots are to be fused into one globally consistent model.
- **Averting Immobilization.** If a robot is immobilized amidst the field owing to complete expenditure of the allocated resources, then, additional robots may need to be deployed to retrieve the stranded agent. Retrieval of the stranded agent is important to obtain the model generated by the agent, but doing so with the help of additional agents will increase the overall project cost and delay further deployments.

8.2.1 Dynamically choosing weights for optimization

Thus far, static weights were chosen for *RC-DAS* acquisition function, but an apt choice drastically modifies the behavior of the robot. This can be attributed to the fact that the selection of the weights biases the robot to either pay heed to model quality or resource utilization, or both. Thus, in what follows, an optimal weight (α) selection mechanism is devised such that the weights are dynamically updated as a function of residual budget. This is done as follows:

$$\alpha^{[t]} \leftarrow \frac{B_{res}^{[t]}}{B}, \tag{8.6}$$

where

$$B_{res}^{[t]} \triangleq B_{res}^{[t-1]} - (C_S^{[t]}(x^+) + C_T^{[t]}(x^-, x^+)). \tag{8.7}$$

In Eq. (8.6), the current *weighting factor* ($\alpha^{[t]}$) is determined based on the current residual budget ($B_{res}^{[t]}$) as defined in Eq. (8.7), where the current residual budget ($B_{res}^{[t]}$) is defined as the difference between the previously available budget $B_{res}^{[t-1]}$ and the cost that was incurred in moving and gathering measurement in the previous time step ($t - 1$) such that $0 < B_{res}^{[t]} \leq B$. The current location of the robot is given by x^- and the next location being evaluated is given by x^+. At $t = 0$, $B_{res}^{[0]} \triangleq B$ such that

$\alpha^{[0]} \leftarrow 1$. When plugging in apt weights (α) from Eq. (8.6) into Eq. (8.3), not only can the *next-best-location* be found, but also explorative and exploitative nature of the objective function is adjusted in accordance with the residual resources.

8.2.2 Additional homing constraints

By adapting the weights of the components as the resources are being depleted, the robot can effectively use the available resources to observe most of the field, however, this still suffers from one major loop-hole: the robot cannot return to its base since it does not account for homing while planning paths. To remedy this, homing is posed as additional constraints while planning the trajectory and this revised cost function will now be addressed as RC-DAS† [8]. It looks as follows:

$$
\hat{\mathbf{x}}_{R^{\dagger}} = \arg\max_{\mathbf{x}^+} \left\{ \alpha \, \mathbb{H}_{\mathbf{y}^+ | \mathbf{y}, \theta} - (1 - \alpha) \ln \|\mathbf{x}^- - \mathbf{x}^+\| \right\},
$$
$$
\text{s.t. } \arg\min_{\mathbf{x}^+} \{C_T(\mathbf{x}^-, Home), C_T(\mathbf{x}^+, Home)\}.
$$

(8.8)

In Eq. (8.8), the additional constraint checks if the robot has enough resources to execute the trajectory to reach the chosen candidate and still return to base. Otherwise, the exploration is immediately terminated and the robot returns to the base right away.

8.3 Experiments

The aforementioned information acquisition functions were empirically evaluated on a significantly large scale spatio-temporal environment monitoring dataset which is publicly available.

8.3.1 Dataset

The **US ozone dataset** was used which includes ozone concentrations (in parts per billion) collected by US Environmental Protection Agency (US-EPA) [7]. In this dataset, the measurements were recorded for several years at 59^2 static monitoring stations across USA, but only one of the years was chosen for evaluation of interpolation performance. For each station, the annual average ozone concentration was assigned as the sample measurement for that station. A high speed robot was simulated wherein the speeds and mechanical capabilities of the robots were assumed befitting the requirements of covering extensive distances within small time spans (of the order of seconds).

---

2 After removing missing entries.

8.3.2 Analysis without homing guarantees

The first set of experiments focuses on analyzing the acquisition functions, viz., *RC-DAS* pegged against *full-DAS* without any homing guarantees. For this, a predetermined value of $\alpha = 0.7$ was chosen for *RC-DAS*. This meant that, irrespective of the residual budget, 70% importance was given to the model quality while only 30% importance to resource utilization. In order to evaluate the path length incurred and model quality attained for this parameter setting, the following experiment setup was designed: A 4-robot decentralized team was considered to explore and generate models of the **US ozone dataset**. To avoid solving the collision avoidance issue, the robots were restrained within arbitrarily demarcated zones and were allowed to explore using *full-DAS*, *NN*, and *RC-DAS* active sensing schemes as shown in Fig. 8.4. As is visually apparent, *NN* trajectory is the longest with small-step increments, while *full-DAS* trajectories are short with long-step increments driving the robots significantly far away. The average model accuracy and path lengths incurred by the team are summarized in Table 8.1.

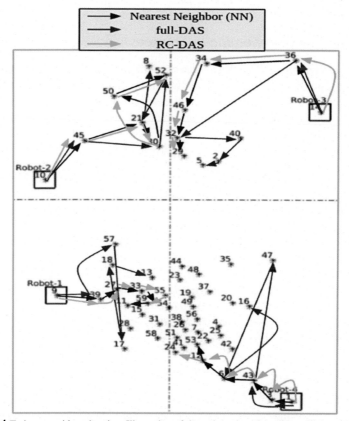

Figure 8.4 Trajectory without homing. Illustration of the trajectories of 4 robots utilizing *full-DAS*, *NN*, and *RC-DAS* active sensing schemes. To avoid collisions, the robots were restrained within preallocated sensing zones as is demarcated with dashed lines. Image taken from [7].

Table 8.1 Analyzing performance without homing.

Item	full-DAS	RC-DAS	NN
Average RMSE	23.671 ± 3.526	$\mathbf{19.3259 \pm 1.820}$	21.128 ± 7.034
Average Path Length (Kms)	9.5 ± 3.879	$\mathbf{16.5 \pm 4.319}$	17.1 ± 4.346

From Table 8.1 it was concluded that the average path cost incurred by *full-DAS* is the least, while that of *NN* is the maximal, which is in accordance with Fig. 8.3. Owing to scanty numbers of highly informative samples, the *full-DAS* model cannot attain optimal performance and neither can *NN*, owing to hoards of correlated observations. Thus, *RC-DAS* is elegantly placed in the middle with an optimal model quality and bounded path length.

It is important to reiterate here that these results were obtained for a fixed $\alpha = 0.7$, but, in reality, a robot will not be able to sustain this highly explorative nature as the resources are being depleted. Thus, in the following, the impact of homing and dynamic weight updates will be evaluated. Also, since *NN* avails no benefits in terms of model quality, only *full-DAS* and $RC\text{-}DAS^{\dagger}$ were evaluated.

8.3.3 Analysis with homing guarantees

For this set of experiments, $RC\text{-}DAS^{\dagger}$ was compared to *full-DAS* whereby the former inherently ensures homing while the latter was artificially enforced to do so. The analysis is split into two segments: *first*, the analysis of trajectories is carried out, followed by the analysis of model quality hence generated.

8.3.3.1 Path cost analysis with homing enforced

Firstly, the length of a walk, i.e., number of locations observed by a single robot starting from multiple start locations is analyzed. The results are shown in Fig. 8.5, wherein each trend represents a random starting location such that each of the available 59 locations was chosen as a start location at least once.

From Fig. 8.5 it can be deduced that $RC\text{-}DAS^{\dagger}$ is highly conservative in utilizing the available resources and hence can allow the robots to observe more locations. The length of a walk of $RC\text{-}DAS^{\dagger}$ is significantly larger than that of *full-DAS*. Besides, this also proves the invariance of the $RC\text{-}DAS^{\dagger}$ to the choice of start location. This is important because the available information guides the robot along the steepest gradient, and if the robot is adversely affected by the choice of the start location then convergence cannot be guaranteed.

8.3.3.2 Model quality analysis with homing enforced

In what follows, the model quality was analyzed when using *full-DAS* and $RC\text{-}DAS^{\dagger}$ to ensure that imposing homing guarantees does not compromise the model quality significantly. Before doing so, a few evaluation criteria are defined in Definitions 8.9 and 8.10.

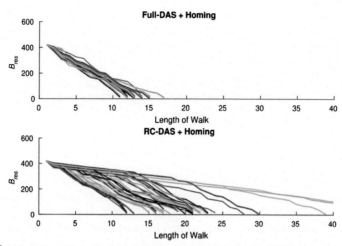

Figure 8.5 Budget decay. Analyzing how the budget is consumed (decayed) while gathering observations using the *full-DAS* and *RC-DAS*† active sensing schemes. Tests are performed for artificially enforced homing constraints using *full-DAS*. Each trend represents budget decay for the respective scheme for a chosen starting point. Image taken from [8].

Definition 8.9 (Model performance). The model performance when using a chosen active sensing scheme is defined as the *Root Mean Squared Error (RMSE)* over the predicted measurements, $\forall x^* \in U$, for a robot. The lower the RMSE, the better the model performance, and hence the more accurate the map.

Definition 8.10 (Precision (P)). If a total of N experiments are performed during which N_f represents the number of times *full-DAS* generated a more accurate map than *RC-DAS*, and N_R represents the number of opposite outcomes, then precision (P) for *full-DAS* is given by

$$P_F \triangleq \frac{N_f}{N},\tag{8.9}$$

and the precision (P) for *RC-DAS*† is given by

$$P_{R^\dagger} \triangleq \frac{N_R}{N}.\tag{8.10}$$

Thus, P represents the chances of generating a *better model*[3] of the environment given the choice of an active sensing scheme. The model accuracy can be evaluated by comparing the predicted measurements with respect to the ground truth values and evaluating the RMSE to associate a scalar value as a uniform performance measure for the model being considered.

In this set of experiments, the precision of *RC-DAS*† was compared against *full-DAS*. All possible start locations were considered and the average performance is

---

[3] Cf. Definition 8.9.

reported in Table 8.2. Since $RC\text{-}DAS^{\dagger}$ always considers homing, two subsets of experiments needed to be performed: first, considering *full-DAS* in its current form, i.e., without homing, and second, by manually enforcing homing on *full-DAS*.

Table 8.2 Impact of homing on the precision of *full-DAS* vs $RC\text{-}DAS^{\dagger}$.

	P_F	$P_R^{\dagger}$
Full-DAS w/o Homing	63.33%	36.67%
Full-DAS with Homing	45%	55%

From Table 8.2 it is evident that *full-DAS* has a higher precision when homing is not performed, but owing to homing constraints the proposed $RC\text{-}DAS^{\dagger}$ has superior performance. Alternatively, this also states that, when homing is a necessary requirement, $RC\text{-}DAS^{\dagger}$ is more robust to the choice of start location assigned to the robot. Since the start locations directly affect the trajectory and the terminal quality of the prediction model, robustness to the choice of the start location is of utmost importance.

8.4 Summary

A novel class of acquisition functions belonging to the *high variance* active acquisition family was discussed that allows the robot to acquire information from the most uncertain areas whilst placing strict bounds on the net path length, guaranteeing homing and not compromising on the model performance significantly. The acquisition function is meant to scale to any size of the team of mobile robots that may be operating in communication devoid environments in a fully decentralized and disconnected fashion. The results show a significant reduction in net path costs whilst attaining similar or even better accuracy as shown by the state-of-the-art acquisition function, viz., *full-DAS*. Besides this, robustness to start locations was also empirically validated, which is of utmost importance when robots set to venture out in unknown environments. Since the information acquisition functions under consideration follow the gradient of information, if the robot starts from the least informative locations (which cannot be known *a priori*), it may drastically affect the model quality but $RC\text{-}DAS^{\dagger}$ can efficiently handle such scenarios.

References

[1] M. Dunbabin, L. Marques, Robots for environmental monitoring: significant advancements and applications, IEEE Robotics & Automation Magazine 19 (1) (2012) 24–39.
[2] T. Wilson, S.B. Williams, Active sample selection in scalar fields exhibiting non-stationary noise with parametric heteroscedastic Gaussian process regression, in: Robotics and Automation (ICRA), 2017 IEEE International Conference on, IEEE, 2017, pp. 6455–6462.

[3] A. Singh, F. Ramos, H.D. Whyte, W.J. Kaiser, Modeling and decision making in spatio-temporal processes for environmental surveillance, in: Robotics and Automation (ICRA), 2010 IEEE International Conference on, IEEE, 2010, pp. 5490–5497.

[4] J. Chen, K.H. Low, Y. Yao, P. Jaillet, Gaussian process decentralized data fusion and active sensing for spatiotemporal traffic modeling and prediction in mobility-on-demand systems, IEEE Transactions on Automation Science and Engineering 12 (3) (2015) 901–921.

[5] A. Krause, A. Singh, C. Guestrin, Near-optimal sensor placements in Gaussian processes: theory, efficient algorithms and empirical studies, Journal of Machine Learning Research 9 (Feb) (2008) 235–284.

[6] N. Waters, Tobler's first law of geography, in: The International Encyclopedia of Geography, 2017.

[7] K. Tiwari, V. Honoré, S. Jeong, N.Y. Chong, M.P. Deisenroth, Resource-constrained decentralized active sensing for multi-robot systems using distributed Gaussian processes, in: 2016 16th International Conference on Control, Automation and Systems (ICCAS), 2016, pp. 13–18.

[8] K. Tiwari, S. Jeong, N.Y. Chong, Multi-UAV resource constrained online monitoring of large-scale spatio-temporal environment with homing guarantee, in: Industrial Electronics Society, IECON 2017-43rd Annual Conference of the IEEE, IEEE, 2017, pp. 5893–5900.

Part III

Mission characterization
How does one define a mission?

Contents

Mission characterization is a crucial aspect of deploying mobile robots for real-world applications. This part describes the problem setup, under which the algorithms presented herewith have been validated. Additionally, it reviews two classes of mission termination criteria for real-world deployments. A brief overview of each of these chapters is provided below.

III.1 Problem formulation

Until now, robot platforms, conventional robotic applications, and non-parametric machine learning models have been introduced. The aim of this chapter then is to formalize the connection between robots, environment monitoring, and GPs. In essence, this chapter connects the dots, i.e., the key building blocks which lay the foundation of a resource-constrained perspective of environment monitoring using multi-robot teams. Further details of each of these components then follow suit in subsequent parts.

III.2 Endurance & energy estimation

In this chapter, recent findings in the domain of endurance and energy estimation are summarized. Endurance estimation is very specific for aerial robots and has been investigated with respect to both fixed winged and rotor-based mechanisms. As opposed to this, energy estimation is applicable for ground vehicles. While each of these approaches is specific to the respective robot type, the end goal for both approaches remains to prolong the mission within the limits of available resources.

III.3　Range estimation

In this chapter, a novel viewpoint on mission characterization is presented. As opposed to all conventional approaches, a more human-friendly notion of range estimation is presented and validated extensively across multiple robot platforms.

Problem formulation
Connecting the dots

9

The more concretely one understands the problem, the more likely the person is to come up with a robust solution.

Dr. Kshitij Tiwari

Contents

Highlights

- Connection between GPs and robots
- Nature of observations that can be acquired
- Starting configuration for robot deployment
- Requirements for model fusion

Erstwhile chapters have presented non-parametric Bayesian methods, viz., GPs, two classes of robotic path planning, viz., CPP and IPP, and described the various robotic platforms that can be potentially used for monitoring various environments. The aim of this chapter is to connect the dots, i.e., define the roles and interplay of each of these components by explicitly describing the operational setup. Additionally, team deployment configuration and sensing scenario encompassing the potential termination conditions are described.

9.1   Relationship between robots and GP

In what follows, each robot is assumed to be a Gaussian Process (GP) expert. What this entails is that the robot is not only capable of deciding where to get the best ob-

Multi-Robot Exploration for Environmental Monitoring. https://doi.org/10.1016/B978-0-12-817607-8.00023-X

servations from, but is also equally capable of using these observations to train the GP models on-board. Owing to practical and hardware limitations of performing on-board GP inference, most of the results were evaluated in a simulated setup, and thus, sufficient resources were assumed as made available to the robot to carry out its mission. Additionally, the area to be monitored was large, and was covered by multiple robots. This naturally rendered multiple GPs, at the end of mission of each robot.

9.2   Scenario

Most of the environmental phenomena have been traditionally monitored using static sensors placed across the target area. Eventually, this gave rise to discrete datasets which were logged in the format of $< Location, Measurement >$, making the data discrete in nature. For most simulated testing, these are the datasets that are used to evaluate the model fit, thus, the training data available for GPs becomes inherently discrete in nature. This, in contrast to the real-world, where the data is continuous, is what is fed to the GP experts to train their respective models.

9.2.1   Starting configuration

As each robot follows the same objective function which is guided by the amount of information that can be accrued from the candidate locations, they could not be started off from the same start location. Thus, for the most part of this book, the robots were randomly allocated to start locations over multiple trials.

9.2.2   Communication strategy

Most of the algorithms were designed for operations under harsh environmental conditions like sub-surface operations in case of marine observations or coal-mine monitoring, etc. In such cases, communication is mostly intermittent or non-existent, so the robots were made independent of their peers. Each robot is an "expert" in the sense that it is a master of its own will and the decisions taken remain unaffected by those of the peers.

9.2.3   Sensing

Owing to the discrete nature of the datasets being used for evaluation, each robot was assumed to be equipped with a suitable interoceptive sensor which helps the agent perceive point-samples only where the robot/sensor is located. Additionally, unless specified otherwise, *RC-DAS* was used for all agents to acquire informative samples by trading off the amount of information to the available resources using a static weight for each component.

9.2.4 Mission termination conditions

Two kinds of termination conditions have been discussed in this book, which primarily pertain to the experimental scenario being *simulated* or *real*.

In case of *simulated* testing, the robots were allocated a resource budget as a function of "feasible" travel distance that they are expected to cover. This has previously been used in Chap. 8. As opposed to this, for *real* experiments, an auto-regressive estimator needs to be used to obtain the true residual budget from the robot as the mission progresses. This is described in a greater depth in the upcoming chapters in this part.

9.2.5 Model fusion

Upon termination of each respective mission, each GP expert has its respective optimal parameter. A GP model is represented in terms of its mean and covariance functions which can be encoded via the hyper-parameters, meaning that there are multiple models to describe the single environmental phenomenon that each team member was primarily observing. Thus, for a human observer to finally decipher or use these models, there is a need to develop a globally consistent model which begs the question:

Who can be trusted amongst the team?

This challenge is addressed in a greater detail in Part IV, which focuses on scaling the team size for efficient monitoring, and this aspect is described especially in Chap. 13.

9.3 Summary

This is a rather brief chapter which gives a preview of the underlying assumptions and design considerations that govern most of the algorithms that follow suit. Previously, GPs, robots, and path planning mechanisms were discussed separately, but this chapter attempted to connect these components to lay the foundation of a coherent system that will be used for carrying out environmental monitoring operations. In the upcoming chapters (Chaps. 10 and 11) of this part, two pragmatic mission termination criteria for real-world deployments are described before moving on to a discussion on multi-robot teams in Part IV.

Endurance & energy estimation
How long can the robot sustain operation?

Time once lost is lost forever. Hence, use it wisely.

Ranal Currie

Contents

Highlights

- Characterizing mission in terms of endurance or power requirements
- Applicable mostly to UAVs in terms of flight time
- Extendable to UGVs in terms of energy consumption

Battery powered aerial vehicles are often utilized for applications like aerial surveying post-disaster or checking the progress of a forest fire. For such applications, it is critical that the endurance (maximal flight time) of the aerial vehicle is considered while planning the missions to avoid complete immobilization amidst the mission. In contrast to aerial vehicles, for ground/marine vehicles, it is rather common to evaluate the energy requirements for specific missions as opposed to endurance. This chapter discusses some of the works in the domain of endurance and energy estimation which primarily focus on the usage of aerial and ground vehicles for a variety of applications, respectively.

10.1   Endurance estimation: the notion

"Endurance" refers to the ability of sustenance when faced with harsh conditions without giving up. In terms of the robots being considered for this book, this would refer to the ability for the robot to sustain operation in the field until the battery completely drains. An intuitive example is shown in Fig. 10.1, wherein athletes have to carefully pace themselves, otherwise they get tired and are unable to complete the race. This is especially true for long-distances like marathons. This chapter looks into similar problems but in a robotics setting.

Figure 10.1 Illustration of endurance estimation via an athletic race-track example. If the athletes do not pace themselves, they would not be able to complete the race as the lack-of-breath would affect their stamina.

Battery-powered unmanned vehicles suffer from major limitations with respect to endurance capabilities,[1] due to the low specific energy of batteries as compared to other energy sources, as well as the fact that, unlike chemical propulsion where the mass is reduced as fuel is burnt, the battery mass is fixed irrespective of its state-of-charge (SoC) [2]. A possible solution to enhancing the endurance of a battery-powered rotorcraft was introduced in [3] wherein the authors proposed subdividing the monolithic battery into multiple smaller-capacity batteries which are sequentially discharged and released. Thus, as time passes the robot's payload gets lighter which reduces the requirement for propulsive power over time. However, releasing the used battery and swapping the power source to a fresh battery while the robot is operating is non-trivial and might lead to an increased payload. Thus, whilst attractive to use in exploration missions due to their hovering and low-speed flight capabilities and high maneuverability, the endurance of such expeditions is significantly constrained [4]. As opposed to this swapping and releasing used batteries of smaller capacities, the authors in [5] propose to model the power consumption on-board the commercial off-the-shelf UAVs. In this work, the authors analytically proposed the power consumed for hovering and empirically evaluated the power consumption profiles for preset maneuvers. The proposed power consumption models were then used to obtain the endurance model for rotorcrafts.

---

[1] Endurance of some of the commercial drones is summarized in Table 10.1.

Table 10.1 Endurance of various commercial drones. Statistics based on [1].

Drone	Flight time (mins)
Short Range (0–300 m)	
Drocon MJX X708W	8
UDI U34W Dragonfly	8
Holy Stone F181W	10
Holy Stone HS300	15
UDI U818Plus	15
Medium Range (300–1000 m)	
MJX Bugs 3	15
Holy Stone HS300	15
Upair One	15
MJX Bugs 2W	18
Traxxas Aton	20
Long Range (1000+) m	
GoPro Karma	20
Parrot Bebop 2	21
Yuneec Typhoon H	25
Autel Robotics X-Star Premium	25
DJI Phantom 4	27
DJI Mavic Pro	27
Parrot ArDrone 2.0	36

In [6], the authors presented an analytical model for accurate estimation of hovering time as a function of on-board battery capacity. In [7], the authors proposed a novel setup for energy management for aerial vehicles during indoor exploration missions. In this work, the authors proposed using a ceiling attachment that the robot can cling to while shutting off its motors for free suspension to maintain the birds-eye view helping preserve the energy. For the actual motion, an endurance estimation framework was proposed that takes into account the motor and propeller setup to determine the thrust–power relationship. Interestingly, the estimation model can also be used for identifying the best battery source for the robot amongst the list of available batteries. The same framework was used to estimate the residual endurance while the robot is suspended from the ceiling attachment.

10.2 Energy estimation: the notion

While endurance estimation is primarily focused on aerial robots, one of the crucial questions to develop autonomous (marine/ground) mobile robotic systems is the energy consumption. This encompasses energy consumption modeling, monitoring, and management all along the mission.

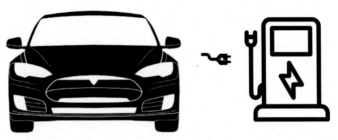

Figure 10.2 Illustration of energy estimation via an electric vehicle recharging example.

Figure 10.3 Nissan Leaf Electic Vehicle (EV). Image courtesy: Kiwiev released under the CC0 License [10].

The ability to predict when power will be depleted beyond a certain threshold is necessary for recharging or returning to a base station. This is the most often used approach, for instance, for (rechargeable) electric vehicles like Tesla, Nio, and automobiles produced by other vendors. A typical recharging setup consists of the vehicle which needs to be physically plugged in to a charging station as shown in Fig. 10.2. Most of the electric vehicles come with a digital (or sometimes physical) gauge array which displays a number of parameters like outside temperature, mileage achieved since last charge, current battery capacity (in %), aside from the conventional odometric information. An illustration of such a gauge for a Nissan Leaf EV is shown in Fig. 10.3. Knowing such metrics is not only essential for electric/hybrid vehicles, but also important for robotics applications so that the mission can be planned aptly. While there are several models available for electric vehicles like [8,9], these methods are not directly applicable to mobile robots owing to significantly different dynamics considerations. Such differences arise from differences in operational environments and significantly different weights of the two.

Hamza and Ayanian in [11] present a framework for forecasting battery state-of-charge (SoC) of a robot for a given mission. A generalized and customizable mission description is formulated as a sequence of parametrized tasks defined for the robot; the missions are then mapped to expected change in SoC by training neural networks on experimental data. Similar attempts were shown in [12] wherein the RoBM^2 board is described, which allows for battery-tailored supervision policies.

With regards to energy management, Dynamic Power Management (DPM) [13] techniques optimize the power consumed by the hardware and software components by dynamically adapting the power consumed based on estimated future requirements. For instance, one such technique is the Dynamic Voltage Stabilization (DVS) which dynamically optimizes the processor frequency and voltage [14]. In [15], the authors empirically identified energy estimation models for two power sources: primary battery which powered the DC motors that control the wheels of the Pioneer 3DX and govern its on-board computational resources, and a secondary battery present on-board the embedded laptop where all the computations were performed and control commands were prepared to be passed to corresponding actuators. In a complementary work [16], the authors discuss the notion of mission "performance" for autonomous mobile robots. In this work, two classes of performance metrics were discussed: *main frame* metrics included safety, energy, localization, and stability, and *user-defined* metrics, namely, mission duration. And the notion was further extended to fault-tolerant autonomous robotic missions in [17].

10.3 Conclusion

Thus far, the researchers either have developed *endurance estimation* models which pertain primarily to aerial robots, or *energy estimation* models which dynamically adapt the energy requirements for components and predict the battery health in terms of state-of-charge, primarily of ground robots. The latter is a generic framework that is applicable to robotic platforms operating in ground or marine environments. However, none of these models can directly answer a simple question:

How far can a robot go whilst operating on a single discharge cycle?

In just a matter of a few words, this "simple" question opens up an entire arena of research encompassing several real-life challenges. These include proposing a novel operational range estimation framework and being able to incorporate current mission statistics into the operational range estimation. Additionally, in a majority of the missions, the robot is not expected or required to move incessantly. Thus, a provision is required for the robot to stop-and-process and still be able to estimate the operational range in the near future. Whilst all these efforts are aimed at avoiding complete battery depletion eventually leading to robot strangulation, *operational range estimation* paradigm is rather pragmatic. Further details are discussed in the following chapter.

References

[1] http://www.dronesfella.com/guide/long-range-drones/.

[2] J. Schömann, Hybrid-Electric Propulsion Systems for Small Unmanned Aircraft, Ph.D. thesis, Technische Universität München, 2014.

[3] A. Abdilla, A. Richards, S. Burrow, Endurance optimisation of battery-powered rotorcraft, in: Conference Towards Autonomous Robotic Systems, Springer, 2015, pp. 1–12.

[4] J.L. Pereira, Hover and Wind-Tunnel Testing of Shrouded Rotors for Improved Micro Air Vehicle Design, Tech. Rep., Maryland Univ. College Park, Dept. of Aerospace Engineering, 2008.

[5] A. Abdilla, A. Richards, S. Burrow, Power and endurance modelling of battery-powered rotorcraft, in: Intelligent Robots and Systems (IROS), 2015 IEEE/RSJ International Conference on, IEEE, 2015, pp. 675–680.

[6] M. Gatti, F. Giulietti, M. Turci, Maximum endurance for battery-powered rotary-wing aircraft, Aerospace Science and Technology 45 (2015) 174–179.

[7] J.F. Roberts, J.C. Zufferey, D. Floreano, Energy management for indoor hovering robots, in: Intelligent Robots and Systems, 2008. IROS 2008. IEEE/RSJ International Conference on, IEEE, 2008, pp. 1242–1247.

[8] J.G. Hayes, R.P.R. De Oliveira, S. Vaughan, M.G. Egan, Simplified electric vehicle power train models and range estimation, in: 2011 IEEE Vehicle Power and Propulsion Conference, IEEE, 2011, pp. 1–5.

[9] C. Bingham, C. Walsh, S. Carroll, Impact of driving characteristics on electric vehicle energy consumption and range, IET Intelligent Transport Systems 6 (1) (2012) 29–35.

[10] https://commons.wikimedia.org/w/index.php?curid=29811585.

[11] A. Hamza, N. Ayanian, Forecasting battery state of charge for robot missions, in: Proceedings of the Symposium on Applied Computing, ACM, 2017, pp. 249–255.

[12] N. Lucas, C. Codrea, T. Hirth, J. Gutierrez, F. Dressler, RoBM2: measurement of battery capacity in mobile robot systems, in: GI/ITG KuVS Fachgespräch Energiebewusste Systeme und Methoden, Erlangen, Germany, 2005, pp. 13–18.

[13] Y. Mei, Y.H. Lu, Y.C. Hu, C.G. Lee, A case study of mobile robot's energy consumption and conservation techniques, in: Advanced Robotics, 2005. ICAR'05. Proceedings. 12th International Conference on, IEEE, 2005, pp. 492–497.

[14] P. Pillai, K.G. Shin, Real-time dynamic voltage scaling for low-power embedded operating systems, in: ACM SIGOPS Operating Systems Review, vol. 35, ACM, 2001, pp. 89–102.

[15] L. Jaiem, S. Druon, L. Lapierre, D. Crestani, A step toward mobile robots autonomy: energy estimation models, in: Conference Towards Autonomous Robotic Systems, Springer, 2016, pp. 177–188.

[16] L. Jaiem, L. Lapierre, K. Godary-Dejean, D. Crestani, Toward performance guarantee for autonomous mobile robotic mission: an approach for hardware and software resources management, in: Conference Towards Autonomous Robotic Systems, Springer, 2016, pp. 189–195.

[17] L. Jaiem, L. Lapierre, K. Godary-Dejean, D. Crestani, Fault tolerant autonomous robots using mission performance guided resources allocation, in: SysTol: Control and Fault-Tolerant Systems, 2016.

Range estimation
How far can the robot go?

If you do not keep track of how far you can go, you cannot return home at the end of the day.

Dr. Kshitij Tiwari

Contents

Highlights

- Contrasting mission characterization in terms of duration versus range
- Motivation for considering range
- Case Study for UGVs
- Case Study for UAVs

Erstwhile, the proposed algorithms for path planning were claimed to be bounded by the "budget", but it is very important to decipher what is meant by this keyword and how the robot's actions lead to dissipation of the budget. Thus, this chapter discusses how to evaluate the maximum attainable range, i.e., operational range for a robot given some prior information about the mission characteristics and the robot itself (including the nature of the power source used to propel the robot). This is what is referred to as "budget" for this work. Accurate estimation of operational range plays a significant role in allowing the robot to plan closed-loop trajectories and prevent complete immobilization amidst the field.

11.1 Importance of Operational Range Estimation (ORangE)

It is a known fact that the K9 unit of the armed forces are highly specialized in enemy takedowns. One such training scenario is shown in Fig. 11.1 wherein the handler is holding on to the leash of the K9 while the other officer playing the role of an adversary (dressed in protective gear) tantalizes the K9. If the handler does not let go of the leash, then the adversary remains safe as long as he does not cross the black dashed line which represents the maximal biting range of the K9 while being held. For the sake of the argument, this discussion ignored the possibility of the K9 being completely let loose to chase down the adversary. In this setting, the maximal biting range that limits the outreach of the K9 is what could be considered as its *operational range*.

Figure 11.1 Armed forces personnel training a K9 officer. The personnel wearing the protective gear remains out of the biting range of the K9 (shown by dashed black line) assuming the handler does not let go of the collar. Image based on [1].

Similar to the K9s, mobile robots are being increasingly deployed to assist in situations where human intervention is deemed risky or tedious, e.g., actively pursuing evaders, patrolling sensitive areas like cross-country borders or high rise buildings (Fig. 11.2), exploring outer space or assisting in search-and-rescue (Fig. 11.3). Mobile robots are also assisting with house-hold chores and mundane delivery tasks. E.g., Fig. 11.4(A) shows a delivery robot out for delivery of food to a local neighborhood

 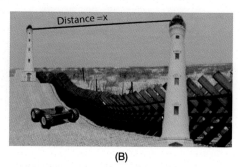

(A)   (B)

Figure 11.2 Operational range estimation for defense applications. (A) shows a multi-level parking lot is shown, which serves as a vantage point for armed forces to provide over-watch when a high-valued individual visits a public spot. These spots once cleared for use by friendly armed personnel cannot be left unmanned. (B) depicts a border patrol scenario is shown where the robot must know its operational range in order to plan the patrolling missions between watch-towers. Image taken from [2].

(A)   (B)

Figure 11.3 Operational range estimation for exploration applications. Robots assisting with government missions. (A), a planetary rover exploring the surface of Mars is shown. (B) shows a ground robot assisting in search-and-rescue after an avalanche is illustrated. In either case, the robot cannot abandon its post or its mission while it is actively surveying. Thus, re-charging is infeasible and the terrain is highly uneven and unpredictable.

while Fig. 11.4(B) shows a vacuum cleaning robot in process of cleaning the household floor. In such time-critical scenarios, the robots cannot abandon their posts for re-charging while carrying out their respective missions.

Most robots have a rough estimate of their battery life based on the operation time [4–8]. Nonetheless, regardless of how the mission is carried out, robots must be retrieved when the operational time is close to the estimated maximum value, or the estimated remaining battery time is close to zero. However, this general approach neglects two facts: (i) missions are highly dynamic, which in turn incurs variable power consumption, thereby making the nominal estimated battery life time too broad, and especially too conservative; (ii) in outdoor/indoor or space exploration missions, other than the mission time, researchers are more concerned about the proportion of the unknown area that the robot covers. Not much research has been done to look into how the energy stored in the battery is distributed amongst different components and how this would affect the maximum traversal range.

Figure 11.4 Operational range estimation for civil applications. (A) Delivery robot called *Kiwibot* out to deliver food; (B) Vacuum cleaning robot in the processing of cleaning the household. Stopping to recharge is infeasible for either scenario and hence the paths must be planned carefully to finish the mission before the battery runs out. Image taken from [3].

Most of the aforementioned works consider the mission profile to be known *a priori* and then try to estimate/optimize the energy requirements based on locomotion models only. As opposed to this, it is more rational to consider the energy itself to be fixed and then modulate the mission profile accordingly. The reason for such an unconventional "inverse" setting is that, given the battery characteristics, the amount of energy stored within and the rate of discharge are known *a priori* and serve as the limiting factor for the mission, when the robots set out to explore the environment. Thus, the inverse problem should be tackled where the energy constraint is known and the path length must be optimized within these constraints. The benefits of the knowledge about the maximal operational range based on a reasonable energy model can be harnessed for both tele-operated robots and fully autonomous robots. This means that either the operators can know the optimal time to retrieve the robots for a tele-operated robot, or, in case of an autonomous robot, the robot itself can gauge the best time to return to its base. Using too much of the provided battery energy could lead to complete failure of the robots amidst the mission and using few resources could significantly reduce the environmental perception.

In this chapter, two important keywords are introduced: **framework** and **model**. Since these cannot be used interchangeably, they have been explicitly defined in Definitions 11.1 and 11.2 in a later part of this chapter.

11.2   Rationale behind the maverick approach

The existing models for endurance [9] or mission energy estimation [10] are made specific to the robot under consideration.[1] Furthermore, the author(s) believe that

---

[1] For a refresher, please refer to Chap. 10.

there exists almost no framework to estimate the operational range of robots functioning on a single discharge cycle, let alone be comprehensive enough to estimate the operational range for any category of robot. This is the philosophy behind the proposed operational range estimation architecture which not only considers the locomotion model along with the ancillary functions which tend to draw power from the same source, but also accommodates any robot while retaining the same composition.

In order for the robot to know its maximal operational range, it is very important that an *energy dissipation model* is developed, which elucidates how the energy from the power source is distributed across the system along with the uncalled-for losses owing to heat dissipation, friction, etc. This chapter takes into account multiple such energy consuming sources and analyzes how the energy stored in the battery is consumed during a single discharge cycle. The research problem addressed here is:

> *Given a robot equipped with a fully charged battery, what is the maximum attainable range on a single discharge cycle to avoid complete immobilization?*

The task at hand is challenging owing to the fact that different robots have distinct motion models and usually operate in diverse environmental conditions which cannot be anticipated in advance. Moreover, making a generic framework that can help estimate operational range for a multitude of robots is useful, yet non-trivial.

11.3 Workflow

For the ease of the readers, this section is meant to provide a roadmap to the sections as they follow hereinafter. The organization of the rest of this chapter is shown in Fig. 11.5 and is detailed as follows: The overall range estimation framework is categorized into 2 types:

- *Simplified Framework* meant for simple environmental conditions focusing only on UGVs is first presented in Sect. 11.4. Within this section, a simplified energy distribution model with corresponding losses is accounted for in Sect. 11.4.1, along with the corresponding range estimation model presented in Sect. 11.4.2.
- *Generalized Framework* which is an extension of the previous framework meant to account for various classes of robots. This is discussed in Sect. 11.5. As before, the enhanced energy distribution model is presented in Sect. 11.5.3, followed by the enhanced range estimation model in Sect. 11.5.4. In order to estimate the range using the generic framework, an *offline* variant is presented in Sect. 11.5.4.1, followed by its *online* counterpart as discussed in Sect. 11.5.4.2. For each of these models, two case studies are presented using UGVs and UAVs in Sects. 1, 2, 3, and 4, respectively.

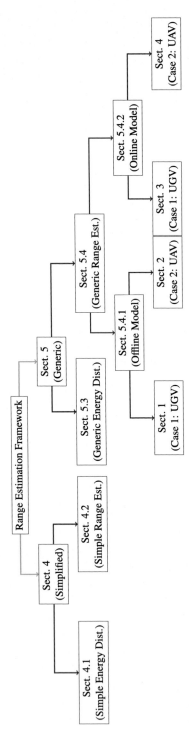

Figure 11.5 Workflow for operational range estimation. Image taken from [2].

11.4 Simplified range estimation framework for UGV

This section presents a simplified framework focusing only on UGVs operating on smooth terrains and smoothly varying elevations like those shown in Fig. 11.2 [11]. A framework is said to comprise two components, viz., (i) energy distribution model which explains how the energy is consumed across the system, and (ii) range estimation model which transforms the useful energy into attainable distance.

11.4.1 Energy distribution model

Consider a rudimentary robot equipped with only a power source and motors for propelling the robot. In such a setting, the energy from the battery is directly used for maneuvering as shown in Fig. 11.6. However, almost never does a robot have such limited payload. A robot that is being used for exploration of unknown environment, be it search-and-rescue or patrolling, always has sensors, micro-controllers, and other essential payload deemed necessary for the respective mission. In such a setting, energy from the battery is distributed amongst the maneuvering and ancillary branches, and also sustains losses as shown in Fig. 11.7.

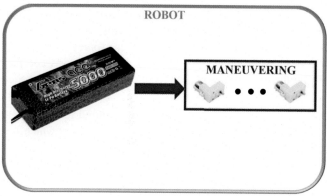

Figure 11.6 Ideal battery dissemination model. All energy stored in a battery is used for maneuvering without any losses. Image taken from [2].

There are four kinds of losses associated with the entire robotic system which, in turn, affect the maximum attainable range of a mobile robot. They are:

- **Battery charge storage loss** (η_1) referring to the battery self-discharge characteristics. Even without any load attached, the battery tends to suffer self-discharge, thereby reducing the net amount of energy available for a mission;
- **Drive motor losses** (η_2) owing to internal friction along with actuation losses;
- **Mechanical losses** (η_3) referring to power train losses like friction in transmission, damping from lubricants, etc.;
- **Ancillary losses** (η_4) accounting for heat losses incurred by sensors, motor drivers, micro-controllers, etc.

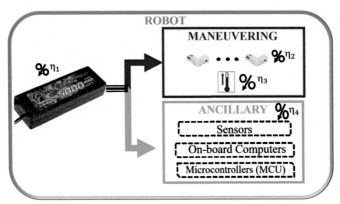

Figure 11.7 Realistic lossy battery dissemination model. Energy from a battery is distributed amongst the maneuvering and ancillary branches. The sustained losses are marked herewith. Image taken from [2].

So, the overall system efficiency can be summarized as $\Omega \triangleq \Pi_{i=1}^{4} \neg \eta_i$ and the procedure for calibration to obtain these losses is explained next.

11.4.1.1   System identification

The procedure of calibrating the system to deduce the four kinds of losses mentioned above is referred to as *System Identification*. For this, a *minimal load test* was designed, in which the only load drawing power from the source were the motors. During this test, the motors were allowed to run until the battery was completely drained and the following three quantities needed to be calculated in order to identify system losses:

- **Internal Friction Power** (P_{IF}) is the power required to overcome the internal friction offered by the motors during the *minimal load test* which lasted for time T_{test}. All the energy supplied from the battery to the motors (E_{motor}) is transformed into mechanical energy (E_{mech}) to turn the motors and motor losses as heat[2] $(I_{motor}^2 R_{motor} T_{test})$ dissipated to the environment. Thus, the following relationship holds from the law of conservation of energy:

$$\underbrace{\tilde{E}_{motor}}_{\text{Net energy for motors}} = \underbrace{E_{mech}}_{\text{Mechanical energy}} + \underbrace{I_{motor}^2 R_{motor} T_{test}}_{\text{Heat loss during test}},$$

$$\Rightarrow P_{IF} = I_{motor}^2 R_{motor} \qquad (11.1)$$

$$= \frac{\tilde{E}_{motor} - E_{mech}}{T_{test}}.$$

- **Field Trial Power** (P_{FT}) is the power consumed during the field trials which lasted for time T_M. Besides the power required to overcome the internal friction of the motors, additional power will now be consumed to overcome terrain resistance

---

[2] Provided by the Joule's first law of heating a.k.a. ohmic/resistive heating.

(P_{terr}). Thus, the overall energy from source is now related to internal friction and terrain interactions as:

$$\underbrace{\tilde{E}_{motor}}_{\text{Net energy for motors}} = \underbrace{E_{mech}}_{\text{Mechanical energy}} + \underbrace{P_{IF}T_M}_{\text{Heat loss during test}} + \underbrace{P_{terr}T_M}_{\text{Terrain interaction}} ,$$

$$\Rightarrow P_{FT} = \frac{P_{IF}T_M + P_{terr}T_M}{T_M}.$$

$$(11.2)$$

- **Net Component Power** (P_{CP}) is the net energy consumed by all components on-board including motors, sensors, controllers, etc., for a mission duration T_M. This is given by:

$$P_{CP} \triangleq \underbrace{\frac{\tilde{E}_{motor}}{T_M}}_{\text{Motor power}} + \underbrace{P_{IF}}_{\substack{\text{Motor loss per unit } T_M}} + \Sigma_{\forall sensor}(\underbrace{P_{sensor}}_{\text{Sensor power}} + \underbrace{I_{sensor}^2 R_{sensor}}_{\text{Sensor loss per unit } T_M}).$$

$$(11.3)$$

Now, the system losses can be calculated for the mission time T_M as:

- $\eta_1 \triangleq 100 \times \frac{\tilde{E}_{net} - P_{CP}T_M}{\tilde{E}_{net}}$ where $\tilde{E}_{net} \triangleq \tilde{E}_{motor} + \Sigma_{\forall sensor} \tilde{E}_{sensor}$;
- $\eta_2 \triangleq 100 \times \frac{\tilde{E}_{motor} - P_{IF}T_M}{\tilde{E}_{motor}} = 100 \times \frac{\tilde{E}_{mech}}{\tilde{E}_{motor}}$;
- $\eta_3 \triangleq 100 \times \frac{P_{FT} - P_{IF}}{P_{FT}}$;
- $\eta_4 \triangleq 100 \times \frac{\Sigma_{\forall sensor}(\tilde{E}_{sensor} - I_{sensor}^2 R_{sensor}T_M)}{\Sigma_{\forall sensor}(\tilde{E}_{sensor})}$.

Here, the zeroth order polynomial has been used, i.e., the first order approximation of the efficiency/losses of the system to estimate its lower bound. However, if a more complex model (higher order polynomial) were to be used that perhaps can also account for mechanical degradation, changes in current demands owing to variable motor loads, elevation changes, and operational velocity modulations, etc., better estimates could be obtained. An even higher complexity model could also track the changes in these parameters in real-time which can be used to account for system efficiencies in an *online* fashion and maintain tighter bounds. Having said this, the challenge still remains to identify such models and quantify their parameters. For the scope of this work, only the first order approximations were retained.

11.4.2 Simplified Operational Range Estimation (ORangE)

Having modeled the energy distributed across the entire system using Fig. 11.7, the maneuvering energy can now be transformed into operational range. For this, the free-body diagram of a robot on an elevated plane (gradient given by γ) is drawn in Fig. 11.8 wherein all the forces in the state of equilibrium are shown, along with the friction offered by the terrain itself which varies based on the surface type and

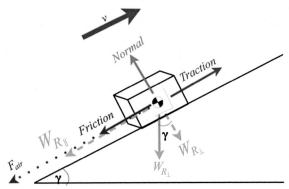

Figure 11.8 Free Body Diagram, illustrating all forces that impede the motion of a robot navigating at an inclination of γ. The cuboid shown represents the robot with all forces acting around its center of mass: $W_R (= m_R g)$ represents the weight of the robot. The weight components were decomposed into the parallel component $W_{R_\parallel} (= m_R g \sin \gamma)$ acting along the terrain and the perpendicular component $W_{R_\perp} (= m_R g \cos \gamma)$ acting against the $Normal(N)$ which represents the normal force. $Friction$ represents the surface friction offered by the ground, and F_{air} represents the aerodynamic drag force. Image taken from [2].

wheel built. Also, the impact of aerodynamic drag force[3] was accounted for. From this figure, the following equilibrium conditions were deduced:

$$N = m_R \ g \cos \gamma \,,$$

$$Traction = Friction + F_{air} + W_{R_\parallel} \,, \qquad (11.4)$$

$$\Rightarrow Traction = C_{rr} \ N + cv^2 + m_R \ g \sin \gamma \,.$$

Thus, in a realistic setup, the energy needed for displacing the robot by an amount d on an elevated terrain can be given by:

$$ME = Traction \times d$$

$$= (C_{rr} \ N + cv^2 + m_R \ g \sin \gamma)d \qquad (11.5)$$

$$= (C_{rr} \ m_R \ g \cos \gamma + cv^2 + m_R \ g \sin \gamma)d \,.$$

In Eq. (11.5), an elevation angle $\gamma \in [0, \gamma_{max}]$ was considered, such that increasing the velocity also increases the ME needed to attain the distance d. However, so far system efficiency parameters are yet to be examined. In Fig. 11.7, two main consumers of the battery energy were presented: *firstly*, the **maneuvering module** which accounts for traversal, steering, etc., and *secondly*, the **ancillary functions module** which accounts for sensing, on-board computations, communication to peers (or, base station), etc. Both these consumers draw power from the same source, which is transformed into useful energy along with unwarranted losses. From [12], the power consumption

---

[3] Although this factor has been considered to make the model realistic, in case of mobile robots, since the operational speed is of the order of a few m/s, this factor can be neglected.

model for sensors was obtained and extended to also account for energy consumed during computations and communications as:

$$P_{anc} = \underbrace{\{s_0 + s_1 f_s\}}_{P_{sensor}} + P_C. \tag{11.6}$$

The advantage of using this power model for the ancillary branch is that it can elegantly take care of situations when the sensor is idling or when it is actively gathering measurements. In Eq. (11.6), the terms in braces refer to the power consumed for gathering measurements, whereby f_s refers to the sampling frequency (Hz) which is contingent on the sensor type, e.g., in case of laser range finders, sonars, and ultrasonic sensors it could refer to the number of rays emitted per second, whilst in case of a camera it could refer to the frames per second (fps) rate. Also, computation cost will only be incurred when sensor measurements are gathered. The term P_C accounts for two factors: (i) the power utilized by micro-controllers to command the wheels and sensors, and (ii) the power used by the on-board computation module. Since the micro-controller tasks are usually fixed, the author assumes a static power consumption, which is stable [12], but the power consumption for computational requirements may vary based on the different tasks like *SLAM, Localization, Occupancy Grid Mapping (OGM)*, etc. Thus, the P_C jointly accounts for the power consumed for computations and short-range wireless communications. However, if more complex models for architectural power consumption were deemed necessary, then the readers are referred to other works like [13].

Motivated by the exponential battery discharge model from [14], this work also uses an exponential decay function to represent this trend using positive coefficients k_1 and k_2 as:

$$\tilde{E} \triangleq E_O \exp^{-(k_1 \zeta + k_2 t)} . \tag{11.7}$$

In Eq. (11.7), as opposed to prior works, the author(s) also considered the impact of several charge–discharge cycles (ζ) along with the age of the battery (t). The coefficients k_1 and k_2 represent the impact of each of these components on the overall battery decay and E_O represents the rated energy available from the battery.

In order to estimate the maximum achievable range, first, the total energy model in a real world setting is established as the sum of the Ancillary Energy (AE) and the Traversal Energy (TE)[4]

---

[4] Net Maneuvering Energy (ME) available, i.e., $\dfrac{ME}{^{UGV}\Omega_{man}}$, where $^{UGV}\Omega_{man}$ is the efficiency of maneuvering branch.

$$\tilde{E} = AE + TE,$$

$$= Ancillary\ Power \times T_M + \frac{ME}{UGV\,\Omega_{man}}$$

$$= P_{anc} \times \frac{d}{vD} + \frac{(C_{rr}\,m_R\,g\cos\gamma + cv^2 + m_R\,g\sin\gamma)d}{UGV\,\Omega_{man}}$$

$$= d \times \left\{ \frac{P_{anc}}{vD} + \frac{(C_{rr}\,m_R\,g\cos\gamma + cv^2 + m_R\,g\sin\gamma)}{UGV\,\Omega_{man}} \right\}, \tag{11.8}$$

$$\Rightarrow d = \frac{\tilde{E}}{\left\{ \dfrac{P_{anc}}{vD} + \dfrac{(C_{rr}\,m_R\,g\cos\gamma + cv^2 + m_R\,g\sin\gamma)}{UGV\,\Omega_{man}} \right\}}.$$

Now, in order to evaluate the theoretical maximum attainable range, the optimal operational velocity (v_{opt}) and reduced battery capacity need to be considered. Thus, the theoretical maximum is given by

$$d_{max} = \left\{ \frac{\tilde{E}}{\dfrac{P_{anc}}{v_{opt}D} + \dfrac{(C_{rr}\,m_R\,g\cos\gamma + cv^2 + m_R\,g\sin\gamma)}{UGV\,\Omega_{man}}} \right\}. \tag{11.9}$$

In Eqs. (11.8) and (11.9), an additional symbol D was utilized, which stands for the *duty cycle*. Albeit the author(s) assume constant operational velocity for carrying out the mission, the robot may sometimes get either overwhelming or too sparse amounts of data (see Fig. 11.9(A)) or may lose connection with the base station (see Fig. 11.9(B)) for which it must stop and manage the situation. To allow the robot to do so, the term D is very important, which represents the proportion of the net mission time which the robot spends for actually moving and covering ground. The term D additionally accounts for the fact that the ancillary power is consumed incessantly throughout the mission and, as the robot stops more often, i.e., $D\downarrow$, the ancillary power (AE) $\uparrow$. Also the theoretical upper bound on operational range, i.e., d_{max} is calculated by using the optimal velocity v_{opt}. The choice of v_{opt} is rather challenging since this is determined by the safe operational velocity given the distribution of obstacles in the environment and environment conditions (nature of terrain, average elevation, etc.). Thus, for the scope of this work, only the safe operational velocity is taken as v_{opt}, and this is determined by the human operator.

In Eq. (11.8), the ancillary power (P_{anc}) is computed with respect to the mission time which is calculated as the ratio of distance to the average speed (i.e., velocity normalized by duty cycle, D; $T_M = \frac{d}{vD}$), while the maneuvering energy is computed with respect to the travel distance, d. Mechanical efficiency (Γ) is the ratio of the energy that is actually used to accomplish mechanical work to propel the robot forward to the total energy that actually goes into the maneuvering branch. It takes into account the aforementioned losses η_2 and η_3: $\Gamma = (1 - \eta_2) * (1 - \eta_3)$; η_1 accounts for the energy loss before the battery output, which is embedded in Eq. (11.7); η_4 is the percentage of

(A)   (B)

Figure 11.9 Need for introducing duty cycle (D). (A) Visual SLAM called ORB-SLAM [15] is shown wherein sometimes not enough features may be available for the robot to localize itself based on the frames captured by the camera(s) and on some other occasions, overwhelming numbers of features may be extracted like in cases of extremely cluttered environments. Image taken from [15]. (B) A common communication channel fault is shown, whereby the robots may occasionally face technical difficulties while parsing messages to and from the base station. In either scenario, the robot needs to wait to recover and can only proceed when the problem has been resolved.

the battery output energy that goes into the ancillary branch: $AE = \eta_4 \tilde{E}$. The overall system efficiency can then be summarized as $\Omega \triangleq \Pi_{i=1}^{4} \neg \eta_i$, where $\neg$ represents the complement operator which is used to obtain the efficiencies from losses.

11.5 Generic Range Estimation (ORangE) framework for diverse classes of robots

The prior work [11] is now extended to not only account for environmental factors like variable and uneven terrains, strong wind gusts, etc., to be faced in real exploration missions shown in Fig. 11.3, but also broaden the outreach of the operational range estimation framework to account for different classes of robots like UGVs and UAVs as shown in Fig. 11.10. Furthermore, *offline* and *online* variants for range estimation and an enhanced ancillary power model are presented, and the empirical performance is discussed henceforth.

11.5.1 First things first

Before presenting the unified framework [17], a.k.a. generic framework, it is essential to clarify the difference between the terms "framework" and "model". They are formally defined in Definitions 11.1 and 11.2, respectively, and these are further aided by Fig. 11.10, wherein "framework" refers to the entire architecture which is applicable to all classes on robots on a high level and "model" refers to inherent modules that model the energy distribution of the system and propose a methodology to transform this into operational range. The only adjustment which must be made as a function

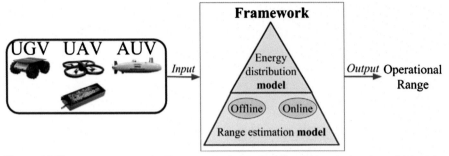

Figure 11.10 Framework v/s Model. **Framework** refers to the entire architecture in unison which can be considered as a black box for operational range estimation. The nature of the robot and its power source are fed as inputs and the operational range is generated as output. **Model** refers to intrinsic components, viz., the energy distribution model for inferring the distribution of energy in the system and range estimation model which convert the useful energy into operational range. On a high level, the black-box can account for various robot types while the differences occur only at the lower levels for adjusting to the distinct locomotion models. Hence, the notion of *unification* is justified. Image taken from [16].

of the robot type occurs at the lower level modules of *offline/online* estimators which take into account the variety of locomotion models. Thus, the notion of *unification* comes into play.

Definition 11.1 (Framework). The pandora's box which takes in the description of the robot and the power source being used in order to deduce the maximum attainable operational range. It only provides a theoretical upper bound on the maximum range and does not include a path planner to assist with homing.

Definition 11.2 (Model). Intrinsic component of framework that serves a specific purpose. In order to deduce the operational range of a robot, several models need to be utilized so that the necessary results can be coupled to deduce the maximum attainable range. The inclusion of all necessary modules gives rise to a framework.

11.5.2 Enhancements over the simplified framework

In this section, the novel additions to the operational range estimation framework are discussed. So far, stand-alone researches have looked into development on analytical models for mission energy and time consumption of specific robotic platforms. Their main focus was to estimate either the endurance or the energy requirements for robots given a pre-set mission. Some models were made *offline* whilst others were generated *online* based on real-time operation data. However, the major limitation of these models was that none of them could estimate the maximal operational range of the robot given some *a priori* known information about the execution of the mission. Furthermore, premeditated trajectories were considered which are not feasible for real-world applications and robot-specific models were developed.

Thus, the aim now is to develop one global framework such that given the battery capacity of a robot which may or may not be further aided by some *a priori* known characteristics of the mission, the maximum operational range for any type of mobile robots can be estimated. Not only this, but a variety of unforeseen environmental factors like sudden changes in terrain elevation or wind gusts, etc., along with the flexibility to stop and process the data are now catered to. These factors inherently affect the maximum attainable range, and avoiding any premeditated trajectories makes the framework better suited to pragmatic applications. Additionally, the ancillary power consumption model from Eq. (11.6) is now further revised to account for variable data traffic using wireless communication. For converting the rest of the useful maneuvering energy into operational range, two variants of range estimation model are also proposed and validated.

11.5.3 Energy distribution model for diverse robots

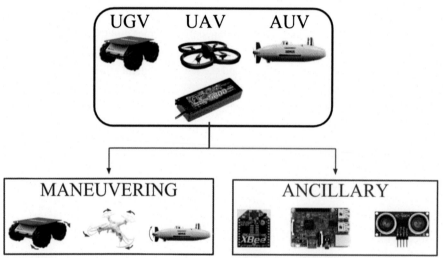

Figure 11.11 Energy distribution model for unification framework. Any type of robot, whether a micro UGV, quadrotor UAV, or AUV, uses portable battery packs which are utilized for essentially two functions: *firstly*, *maneuvering* like propulsion, hovering, navigation, etc., and *secondly*, *ancillary functions* like wireless communication, sensing, on-board processing, etc. Image taken from [16].

The upgraded energy distribution model of the unified framework is shown in Fig. 11.11. Irrespective of the nature of the robot, the energy available from the battery is always utilized by two kinds of processes, viz., *maneuvering* and *ancillary functions*. The proportion of the energy used for the former is referred to as *traversal energy* (TE) while that of the latter is referred to as ancillary energy (AE). In an ideal situation, the net (reduced) energy from the battery $\tilde{E}$ is related to the *traversal energy* (TE) and the ancillary energy (AE) as

$$\tilde{E} = AE + TE. \tag{11.10}$$

Based on the system losses as previously discussed in Sect. 11.4.1, the overall system efficiency of any robot (r) can be summarized as $^r\Omega \triangleq \Pi_{i=1}^{4} \neg\eta_i$. The maneuvering efficiency is given by $^r\Omega_{man} \triangleq \Pi_{i=2}^{3} \neg\eta_i$ and the ancillary efficiency is given by $^r\Omega_{anc} \triangleq \neg\eta_4$.

As for the *traversal energy*, any robot (r) carrying out a mission (m) in an environment of choice experiences four kinds of forces:

1. Constant resistive force $F(r, m)$ as a function of robot (r) and the mission (m), e.g., the force acting on a robot when it is traversing in a straight line under the influence of a constant magnetic field.
2. Environment dependent force $F(x, r, m)$ which is dependent on the current position x, e.g., changing gravitational potential along with changing frictional force because of change in coefficient of friction.
3. Time dependent resistive force $F(t, r, m)$ which is a function of current time t, e.g., unforeseeable disturbances (strong wind gusts, etc.).
4. Instantaneous operational velocity dependent resistive force $F(v, r, m)$ which varies with instantaneous velocity v, e.g., aerodynamics and gyro effect.

Thus, the net *traversal energy (TE)* is given in terms of *mechanical energy (ME)* from the longitudinal dynamics model and the net mechanical efficiency $(^r\Omega_{man})$ as:

$$
\begin{aligned}
TE &= \frac{ME}{^r\Omega_{man}} \\
&= \frac{path \int F_{net}dx}{^r\Omega_{man}} \\
&= \frac{path \int \{F(r, m) + F(x, r, m) + F(t, r, m) + F(v, r, m)\}dx}{^r\Omega_{man}}.
\end{aligned}
\tag{11.11}
$$

Then, the instantaneous time (t) can be expressed as a function of position (x), velocity (v), mission (m), and duty cycle (D) as

$$
t = g(x, v, D, m).
\tag{11.12}
$$

During the mission (m), the robot traverses at an instantaneous velocity (v) and a fixed duty cycle (D).[5] Thus,

$$
\begin{aligned}
TE &= \frac{path \int \{F(r, m) + F(x, r, m) + F(g(x, v, D, m), r, m) + F(v, r, m)\}dx}{^r\Omega_{man}} \\
&= \frac{\{F(r, m) + F(v, r, m)\}d}{^r\Omega_{man}} + \frac{d_{path} \int \{F(x, r, m) + F(x, v, D, r, m)\}dx}{d\,^r\Omega_{man}}.
\end{aligned}
\tag{11.13}
$$

---

[5] The readers are hereby cautioned about a slight abuse of notation. This notation of the duty cycle (D) is similar to the notation of the input data tuple (D) as used in Chap. 6 but is being used in completely different context.

Moreover, the ancillary energy (AE) is given by

$$AE = \frac{P_{anc}d}{v_{avg}D},$$ (11.14)

where

$$P_{anc} = \underbrace{\{s_0 + s_1 f_s\}}_{P_{sensor}} + \underbrace{\{P_{computation} + P_{communication}\}}_{P_C}$$
$$= \{s_0 + s_1 f_s\} + \{P_{computation} + k \times data\ size \times f_{comm}\}$$ (11.15)
$$= \{s_0 + s_1 f_s\} + \{P_{computation} + k \times data\ rate\}.$$

As an enhancement to the ancillary power consumption model over Eq. (11.6), here the communication power ($P_{communication}$) is related with both the size of the data and the frequency (f_{comm}) at which the communication takes place. These two terms could be unified into data rate, i.e., the amount of data sent in unit time. The communication power is then proportional to the data rate with a constant coefficient k (cf. [18]), while the computation power ($P_{computation}$) is a function of the task allocated to the robot. A thorough explanation of the task allocation method is beyond the scope of this book. As for the power consumed by sensors given by P_{sensor}, it can be modeled as a function of the sampling frequency f_s. The scalars s_0, s_1 refer to the static power consumption and operational power consumption coefficient, respectively.

Therefore, the operational range for any robot can be generalized to

$$d = \frac{\tilde{E}}{\dfrac{\{F(r,m) + F(v,r,m)\}}{^r\Omega_{man}} + \dfrac{_{path}\int\{F(x,r,m) + F(\frac{x}{vD},r,m)\}dx}{d\,^r\Omega_{man}} + \dfrac{P_{anc}}{v_{avg}D}}.$$ (11.16)

Here, $^r\Omega_{man}$ is the net maneuvering efficiency of the robot, i.e., the percentage of energy used to do actual mechanical work from the maneuvering branch. From Eq. (11.16), it is evident that in order to estimate the operational range, the term $\dfrac{_{path}\int\{F(x,r,m) + F(\frac{x}{vD},r,m)\}dx}{d\,^r\Omega_{man}}$ needs to be approximated and the operational range estimate would be as good as the approximation of this term.

11.5.4 Range estimation models for diverse robots

In order to approximate Eq. (11.16), two different approaches, viz., (i) *offline* estimation which is a one-shot prediction model wherein range estimate is conjectured at the beginning of the mission itself and predictions are not corrected based on the new data acquired during mission, and (ii) *online* estimation whereby the estimation is recursively updated using all available operational data. In the conventional setting, *offline* estimates are generated once all the data is made available, while the *online* estimates are limited to the data currently available. As opposed to this setting, the *offline* model being referred to here relies on defining the required parameter values *a priori* and retaining the estimates. The *online* model, on the other hand, recursively updates the estimates as more data is made available. Furthermore, for each approach, as case studies, particular models for UGVs and multi-rotor UAVs are discussed.

11.5.4.1 Approach 1: Offline Operational Range Estimation (Offline ORangE) for diverse mobile robot platforms

Assuming some *a priori* known mission characteristics like driving profile, terrain attributes and system efficiency, *firstly* an offline model is explained to estimate the operational range for diverse robots.

1 Case 1: UGV operating on uneven terrain

In Sect. 11.4.2, the operational environment was assumed to be a smooth terrain with a constant elevation. However, in reality, this assumption is often rendered invalid as can be seen from Fig. 11.3. To suit such settings, the framework was further updated to envelop uneven terrains with variable elevation making it better suited to real-world scenarios.

Any natural terrain can be modeled using three features: (i) *flats* – smooth surfaces with negligible gradient, (ii) *slopes* – smooth surfaces with appreciable gradient, and (iii) *rubble* – uneven rough surfaces with no particular gradient characteristics. The operational terrain may have an average slope (γ) with respect to which the operational range d should be calculated. In Fig. 11.12, the dashed line represents the actual terrain which must be traversed where QP represents the actual d; QR is the horizontal reference with respect to which the instantaneous road elevation is calculated.

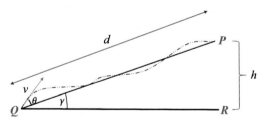

Figure 11.12 Schematic of actual terrain profile without rubble; v represents the instantaneous velocity, γ is the average terrain elevation, and θ represents average road gradient with respect to γ; h represents the elevation gain and d represents the operational range. Image taken from [16].

1. Considering *flat* terrains exclusively, the only resistive force acting on the robot is the (rolling) friction between the wheels and the ground. This is defined as

$$F_{Flats} = Normal\ Force \times C(x)_{rr},\qquad(11.17)$$

   where $C(x)_{rr}$ refers to the coefficient of rolling resistance.

2. Accounting for *slopes*, the net force acting will be friction, together with the weight component, along the motion of the robot. These forces are a function

of the robot location x and the terrain elevation at x given by $\theta(x)$[6]:

$$
\begin{aligned}
F_{slopes} &= F_{flats} + m_{Rg}\sin(\gamma + \theta(x)) \\
&= C(x)_{rr}m_{Rg}\cos(\gamma + \theta(x)) + m_{Rg}\sin(\gamma + \theta(x)).
\end{aligned} \tag{11.18}
$$

3. Finally, considering *rubble*, in the force model, excess forces (F_{rubble}) acting due to presence of rubble need to be accounted for. Let $k(x)_{terr}$ be the terrain coefficient which depends on size, shape, density, and resistance offered by the rubble. Then, the net forces (F_{net}) acting on the robot can be given by

$$
F_{net} = F_{slopes} + F_{rubble}. \tag{11.19}
$$

Considering the limiting case, the following relationship holds:

$$
\begin{aligned}
F_{net} &= F_{slopes} + F_{rubble} \\
&\overset{\Delta}{=} k(x)_{terr}(F_{slopes}).
\end{aligned} \tag{11.20}
$$

Thus, the net maneuvering force ($^{UGV}F(x)_{man}$) for any UGV on an uneven natural terrain is given by

$$
^{UGV}F(x)_{man} = k(x)_{terr}m_{Rg}[C(x)_{rr}\cos(\gamma + \theta(x)) + \sin(\gamma + \theta(x))]. \tag{11.21}
$$

In order to estimate achievable range d, first, the total energy model is defined in a real-world setting as a sum of Ancillary Energy (AE) and Traversal Energy (TE):

$$
\tilde{E} = AE + TE
$$

$$
= \{Ancillary\ Power \times time\} + \frac{\int_{path} {}^{UGV}F(x)_{man}dx}{^{UGV}\Omega_{man}}
$$

$$
= P_{anc} \times \frac{d}{v_{avg}\cos(\theta_{avg})D}
$$

$$
+ \frac{\int_{path} k(x)_{terr}m_{Rg}[C(x)_{rr}\cos(\gamma + \theta(x)) + \sin(\gamma + \theta(x))]dx}{^{UGV}\Omega_{man}}
$$

$$
= P_{anc} \times \frac{d}{v_{avg}\cos(\theta_{avg})D}
$$

$$
+ \left\{ \frac{m_{Rg}}{^{UGV}\Omega_{man}} \times \frac{\int_{path} k(x)_{terr}[C(x)_{rr}\cos(\gamma + \theta(x)) + \sin(\gamma + \theta(x))]dx \times d}{d} \right\}
$$

$$
= d \times \left\{ \frac{P_{anc}}{v_{avg}\cos(\theta_{avg})D} + \frac{m_{Rg}}{^{UGV}\Omega_{man}} \times \frac{\int_{path} k(x)_{terr}[C(x)_{rr}\cos(\gamma + \theta(x)) + \sin(\gamma + \theta(x))]dx}{d} \right\},
$$

$$
\Rightarrow d = \frac{\tilde{E}}{\left\{ \frac{P_{anc}}{v_{avg}\cos(\theta_{avg})D} + \frac{m_{Rg}}{^{UGV}\Omega_{man}} \times \frac{\int_{path} k(x)_{terr}[C(x)_{rr}\cos(\gamma + \theta(x)) + \sin(\gamma + \theta(x))]dx}{d} \right\}}. \tag{11.22}
$$

---

[6] The readers are hereby cautioned not to confuse the terrain elevation $\theta(x)$ with the set of GP hyperparameters represented by $\boldsymbol{\theta}$, cf. Chap. 6.

The maximum attainable distance d_{max} is a function of the optimal velocity v_{opt}. Cruising at speeds higher/lower than v_{opt} would results in operational ranges lesser than d_{max}. Therefore, Eq. (11.22) can now be written as

$$d_{max} = \frac{\tilde{E}}{\frac{P_{anc}}{v_{opt}\cos(\theta_{avg})D} + \left\{ \frac{m_R g}{UGV \Omega_{man}} \times \frac{\int_{path} k(x)_{terr}[C(x)_{rr}\cos(\gamma + \theta(x)) + \sin(\gamma + \theta(x))]dx}{d_{max}} \right\}}.$$

(11.23)

In Eq. (11.23), the factor $\left\{ \frac{m_R g}{UGV \Omega_{man}} \times \frac{\int_{path} k(x)_{terr}[C(x)_{rr}\cos(\gamma + \theta(x)) + \sin(\gamma + \theta(x))]dx}{d_{max}} \right\}$ denotes the average resistive force which acts on the robot on the path QP as shown in Fig. 11.12. Thus, replacing this factor by the expected average resistive force, the maximum traversal range (d_{max}) can be inferred and the estimation accuracy will attain the perfection in the estimation of the expected average resistive force. As the mechanical efficacy of the actuators ($\neg\eta_2$) varies with operational speed, v_{opt} is the velocity at which the net losses of ancillary and maneuvering branches are minimal. Also, as v_{opt} is a rather complex function of robot/actuators, exact trajectory traversed or path taken and the mission characteristics, no further comments or profiling of v_{opt} is possible in the scope of the book. Thus, the target velocity set by the operator was considered as the v_{opt}.

In realistic scenarios, the *offline* model needs the values of $k(x)_{terr}$ and $C(x)_{rr}$ to be defined for each x to find the average expected resistive force. Or equivalently, the integral over the path can be eliminated by replacing it with the average expected resistive force which can be done by replacing $k(x)_{terr}$, $C(x)_{rr}$ and $\theta(x)$ by their averages $\bar{k}_{terr}$, $\bar{C}_{rr}$ and $\bar{\theta}$, respectively. These values for the *offline* model can be estimated using any of the following methods: (i) using the data and experience acquired over the previous missions; (ii) carrying out a trial mission and then using the acquired information as prior knowledge for the actual mission; and (iii) using the expertise of the operators (system/environment experts) to provide realistic/good estimates. For this work, approach (ii) mentioned above was considered.

2 Case 2: multi-rotor UAV operating in the presence of external disturbances

Albeit the energy distribution for a UAV is quite similar to that of a UGV as mentioned previously in Sect. 11.5.3, there are slight variations. The difference with respect to the latter being that, during the mission, a UGV may have phases of negligible maneuvering energy requirements whilst a UAV continuously needs to hover and maintain flight stability. As opposed to [19], the author not only considers the hovering and aerodynamic drag losses, but also accounts for flight adjustments required due to unpredictable environmental factors (like strong wind gusts, etc.).

Analogous to the UGVs, the energy for hovering, drag losses, and flight adjustments in UAVs are comparable to energy requirements of motion over flats, slopes, and rubble, respectively. This is owing to the fact that, in case of hovering, the UAV experiences a constant environment dependent force required to stay aloft. Similarly, to maintain motion for a UGV, it must constantly overcome the resistive frictional forces. Identical analogues can also be drawn for the remaining cases.

1. For *hovering*, the energy consumption model is motivated by the work in [20]. For a UAV with N_R propellers each of whose radii is r_p with a figure of merit Γ and rotor thrust T_{hover}, the power consumed while hovering (P_{hover}) can be defined as

$$
\begin{aligned}
P_{hover} &= \frac{(T_{hover})^{\frac{3}{2}}}{\Gamma r_p \sqrt{2 N_R \rho \pi}} \\
&= \frac{(m_R g)^{\frac{3}{2}}}{\Gamma r_p \sqrt{2 N_R \rho \pi}}.
\end{aligned}
\tag{11.24}
$$

2. Also accounting for *flight adjustments*,[7] the instantaneous power ($P(t)_{fa}$) is given by

$$
P(t)_{fa} = \frac{\left[T(t)_{fa}\right]^{\frac{3}{2}}}{\Gamma r_p \sqrt{2 N_R \rho \pi}},
\tag{11.25}
$$

where the instantaneous thrust with flight adjustments ($T(t)_{fa}$) is defined as

$$
\begin{aligned}
T(t)_{fa} &\triangleq T_{hover} + T(t)_{adjust}, \\
T(t)_{adjust} &= f(t) T_{controller}.
\end{aligned}
\tag{11.26}
$$

In Eq. (11.26), $T_{fa}(t)$ refers to the net thrust required for hovering with adjustments. This is defined in terms of hovering thrust (T_{hover}) and adjustment thrust ($T_{adjust}(t)$). The term $T_{controller}$ refers to the thrust required to follow the acceleration profile generated by the chosen flight controller (e.g., PID controller or Neural Networks, etc.) and $f(t)$ is a time dependent constant of proportionality. In order to expand the outreach of the model and remove the dependence on any particular flight controller, T_{adjust} was modeled as a time dependent function of T_{hover} as

$$
T(t)_{adjust} \triangleq k(t)_{env} T_{hover}.
\tag{11.27}
$$

Thus,

$$
\begin{aligned}
T(t)_{fa} &= T_{hover} + k(t)_{env} T_{hover} \\
&= m_R g + k(t)_{env} m_R g,
\end{aligned}
\tag{11.28}
$$

$$
P(t)_{fa} = \frac{\left[m_R g + k(t)_{env} m_R g\right]^{\frac{3}{2}}}{\Gamma r_p \sqrt{2 N_R \rho \pi}}.
\tag{11.29}
$$

---

[7] The term *flight adjustments* takes into account all adjustments the UAV needs to make in order to maintain its course in the presence of external disturbance or otherwise.

Thus, the average power for *flight adjustments* over the entire time of flight (TOF) is given by

$$P_{fa} \triangleq \frac{\int_{TOF} \left[\{m_R g + k(t)_{env} m_R g\} dt\right]^{\frac{3}{2}}}{TOF \; \Gamma r_p \sqrt{2N_R \rho \pi}}. \tag{11.30}$$

3. Finally, *drag losses* acting on rotor blades need to be incorporated. The drag force on N_R propellers is estimated from fluid mechanics as

$$
\begin{aligned}
F_D &= \frac{N_R \rho C_D A v^2}{2} \\
&= \frac{N_R \rho C_D A (r_p \omega)^2}{2}.
\end{aligned} \tag{11.31}
$$

The drag torque (τ_D) is given by

$$
\begin{aligned}
\tau_D &= \int_0^{r_p} F_D dr \\
&= \frac{N_R \rho C_D A r_p^3 \omega^2}{6}.
\end{aligned} \tag{11.32}
$$

Since the power required to overcome the drag losses is given by $\tau_D \omega$ and the drag thrust $T(t)_{fa}(= N_R k_r \omega^2)$ for propeller constant k_r, the instantaneous power for drag losses $(P_D(t))$ is computed as

$$P_D(t) = \frac{\rho C_D A r_p^3 [T(t)_{fa}]^{\frac{3}{2}}}{6k_r \sqrt{N_R}}. \tag{11.33}$$

Substituting Eq. (11.28) into Eq. (11.33), the average power for drag losses is given by

$$P_D = \frac{\rho C_D A r_p^3 [\int_{TOF} \{m_R g + k(t)_{env} m_R g\} dt]^{\frac{3}{2}}}{6k_r \sqrt{N_R} \; TOF}. \tag{11.34}$$

The net energy required for navigation of a UAV is now given based on Eq. (11.10) as

$$
\begin{aligned}
\tilde{E} &= AE + TE \\
&= Ancillary\ Power \times TOF + \frac{[P_D + P_{fa}]TOF}{UAV\,\Omega_{man}} \\
&= P_{anc} \times TOF + \frac{\left[\frac{\rho C_D A r_p^3 [\int_{TOF}\{m_R g + k(t)_{env} m_R g\}dt]^{\frac{3}{2}}}{6k_r\,TOF\sqrt{N_R}} + \frac{\int_{TOF}[\{m_R g + k(t)_{env} m_R g\}dt]^{\frac{3}{2}}}{TOF\,\Gamma r_p\sqrt{2N_R \rho \pi}}\right]}{UAV\,\Omega_{man}} \\
&= P_{anc} \times TOF + \left\{\frac{\left[\frac{\rho C_D A r_p^3}{6k_r\sqrt{N_R}}\right]+\left[\frac{1}{\Gamma r_p\sqrt{2N_R \rho \pi}}\right]}{UAV\,\Omega_{man}}\right\}\left\{\frac{\int_{TOF}[\{m_R g + k(t)_{env} m_R g\}dt]^{\frac{3}{2}}}{TOF}\right\} TOF,
\end{aligned}
$$

$$(11.35)$$

$$
\Rightarrow TOF = \frac{\tilde{E}}{P_{anc} + \left\{\dfrac{\left[\frac{\rho C_D A r_p^3}{6k_r\sqrt{N_R}}\right]+\left[\frac{1}{\Gamma r_p\sqrt{2N_R \rho \pi}}\right]}{UAV\,\Omega_{man}}\right\}\left\{\dfrac{\int_{TOF}[\{m_R g + k(t)_{env} m_R g\}dt]^{\frac{3}{2}}}{TOF}\right\}}.
$$

$$(11.36)$$

Replacing $TOF = \frac{d}{v}$ in Eq. (11.36), the revised equation becomes

$$
\frac{d}{v} = \frac{\tilde{E}}{P_{anc} + \left\{\dfrac{\left[\frac{\rho C_D A r_p^3}{6k_r\sqrt{N_R}}\right]+\left[\frac{1}{\Gamma r_p\sqrt{2N_R \rho \pi}}\right]}{UAV\,\Omega_{man}}\right\}\left\{\dfrac{\int_{TOF}[\{m_R g + k(t)_{env} m_R g\}dt]^{\frac{3}{2}}}{TOF}\right\}},
$$

$$
\Rightarrow d = \frac{\tilde{E}}{\dfrac{P_{anc}}{vD} + \left\{\dfrac{\left[\frac{\rho C_D A r_p^3}{6k_r\sqrt{N_R}}\right]+\left[\frac{1}{r_p\sqrt{2N_R \rho \pi}}\right]}{UAV\,\Omega_{man} v}\right\}\left\{\dfrac{\int_{TOF}[\{m_R g + k(t)_{env} m_R g\}dt]^{\frac{3}{2}}}{TOF}\right\}}.
$$

$$(11.37)$$

The theoretical maximal operational is only attainable when the UAV operates at v_{opt} such that minimal losses are accrued. Thus,

$$d_{max} = \frac{\tilde{E}}{\left\{ \dfrac{P_{anc}}{v_{opt} D} + \left\{ \dfrac{\left[\dfrac{\rho C_D A r_p^3}{6 k_r \sqrt{N_R}} \right] + \left[\dfrac{1}{r_p \sqrt{2 N_R \rho \pi}} \right]}{^{UAV} \Omega_{man} v_{opt}} \right\} \left\{ \dfrac{\int_{TOF} \left[\{ m_R g + k(t)_{env} m_R g \} dt \right]^{\frac{3}{2}}}{TOF} \right\} \right\}},$$

$\because D = 100\%$ for UAV,

$$d_{max} = \frac{\tilde{E}}{\left\{ \dfrac{P_{anc}}{v_{opt}} + \left\{ \dfrac{\left[\dfrac{\rho C_D A r_p^3}{6 k_r \sqrt{N_R}} \right] + \left[\dfrac{1}{r_p \sqrt{2 N_R \rho \pi}} \right]}{^{UAV} \Omega_{man}} \right\} \left\{ \dfrac{\int_{TOF} \left[\{ m_R g + k(t)_{env} m_R g \} dt \right]^{\frac{3}{2}}}{TOF \, v_{opt}} \right\} \right\}}.$$

$$(11.38)$$

In Eq. (11.38), the factor

$$\left\{ \dfrac{\left[\dfrac{\rho C_D A r_p^3}{6 k_r \sqrt{N_R}} \right] + \left[\dfrac{1}{\Gamma r_p \sqrt{2 N_R \rho \pi}} \right]}{^{UAV} \Omega_{man}} \right\} \left\{ \dfrac{\int_{TOF} \left[\{ m_R g + k(t)_{env} m_R g \} dt \right]^{\frac{3}{2}}}{TOF \, v_{opt}} \right\}$$

represents the average resistive force experienced by the UAV over the entire time of flight. This is akin to Eq. (11.23) which serves to satiate the requirement for developing a *unified framework*. Apt replacement of this parameter by using the expected average resistive force can help estimate the maximum operational range for the UAV. The error in estimation of this factor directly translates to the error in expected operational range.

11.5.4.2 Approach 2: Online Operational Range Estimation (Online ORangE) for diverse mobile robot platforms

In Sect. 11.5.4.1, the author presented an *offline* range estimation model whereby, based on *a priori* known mission characteristics, the maximum attainable range for mobile robots was estimated. However, in reality, it might be rather challenging to strictly follow the mission characteristics or to even obtain *a priori* mission information. Furthermore, the approximation of factors in Eqs. (11.23) and (11.38) are largely dependent on the expertise of the human operator supervising the mission. In order to adapt to unforeseen and unavoidable variations in the mission profile, the author now proposes an *online* variant of the operational range estimation framework. In this method, based on all available real-time data (historic and current), the operational range is recursively updated. As the mission progresses and more data is acquired about the mission characteristics, the range estimation model updates the estimate of the maximum operational range in real-time. This is crucial for real-world missions as it allows for flexibility in missions themselves and can be coupled with energy efficiency path planners [12] to dynamically adapt to situations as they present themselves. Similar to Sect. 11.5.4.1, UGV and multi-rotor case studies are re-introduced in an *online* data acquisition setting.

3 Case 1: UGV operating on uneven terrain

In Eq. (11.23), the terms $k(x)_{terr}$ and $\theta(x)$ can either be set by a human operator (offline model) or can be deduced from prior missions carried out in that terrain. However, the former is prone to human error and the latter is usually not available. Thus, as an alternative, $k(x)_{terr}$ and $\theta(x)$ can instead be replaced by their respective estimates, $\hat{k}(x)_{terr}$ and $\hat{\theta}(x)$. Additionally, the *offline* model used instantaneous rolling resistance $C_{rr}(x)$ while here it is being approximated by a constant C_{rr}. Therefore, the estimated maximum range for the remaining mission is now given by

$$\hat{d}_{Max}^{[t:end]} \triangleq \frac{\tilde{E}_{rem}}{\dfrac{P_{anc}}{v_{opt}D} + \dfrac{\hat{k}(x)_{terr}[C_{rr}\cos\hat{\theta}(x) + \sin\hat{\theta}(x)]m_R g}{^{UGV}\Omega_{man}}}, \qquad (11.39)$$

where

$$\tilde{E}_{rem} = \tilde{E}^{[0:end]} - \tilde{E}^{[0:t]}. \qquad (11.40)$$

Here, $\tilde{E}_{rem}$ is the useful energy remaining in the battery and $\tilde{E}^{[0:end]}$ is the usable energy present in the battery at the start of the mission, i.e., at $t = 0$. Similarly, $\tilde{E}^{[0:t]}$ is the energy spent from $t = 0$ to the time instance t. Now, the total estimated maximum operational range over the entire mission is given by

$$\hat{d}_{max}^{[0:end]} = d^{[0:t]} + \hat{d}_{max}^{[t:end]}. \qquad (11.41)$$

In Eq. (11.41), to estimate the net operational range for the entire mission ($\hat{d}_{max}^{[0:end]}$), the distance that has already been covered ($d^{[0:t]}$) is utilized to estimate the maximum distance that may be covered ($\hat{d}_{max}^{[t:end]}$) based on available residual energy. The value of $\hat{k}(x)_{Terr}$ that is required for estimating $\hat{d}_{max}^{[t:end]}$ can be estimated using Eq. (11.39). During the mission, after every time-step (t), the robot will have the knowledge of the distance that it has covered in that time-step, energy it has spent to cover that distance and terrain elevation θ for that time step. Substituting the value of $d^{[t-1:t]}$ for d_{max} and $E^{[t-1:t]}$ for $\tilde{E}_{rem}$ in Eq. (11.39), the value of $k(x)_{terr}^{[t]}$ for the given time-step t can be calculated. Now, these set of values of $k(x)_{terr}^{[t]}$ can be used to estimate $\hat{k}(x)_{terr}$ which in turn can be used to make predictions about the distance the robot can still cover using the remaining energy. Since the estimate for the remaining distance depends on the value on the estimation of $\hat{k}(x)_{terr}$ and $\hat{\theta}(x)$, which needs to be recursively updated as new data is being collected, the recursive average filter was used. For ease of notation, let $\hat{\mathbf{X}}^{[t]} = [\hat{k}(x)_{terr}^{[t]}, \hat{\theta}(x)^{[t]}]$ and $\mathbf{Z}^{[t-1]} = [k(x)_{terr}^{[t-1]}, \theta(x)^{[t-1]}]$, which represents the set of actual (noisy) measurements up till the last time step ($t - 1$).

Then, given a noisy set of measurements, $\mathbf{Z}^{[0:t-1]}$, and no additional information about the impact of environmental factors on the system dynamics, a reasonable estimate for the system state at the current time-step, t, can be obtained as

$$\hat{\mathbf{X}}^{[t]} = P^{[t-1]} \sum_{i=0}^{t-1} \mathbf{Z}^{[i]} \tag{11.42}$$

where $P^{[t-1]} = \dfrac{1}{t-1}$ represents the responsiveness of the filter, i.e., the filter is very responsive (making a lot of corrections) in the beginning since limited data is available. As time passes and more data becomes available, the filter becomes more certain about its estimates, and thus reduces the relative importance of the measurements. However, being a fixed response model (true values of k_{terr} and θ_{terr} are fixed) with response rate decreasing with time, it cannot always adapt to sudden changes in the values of $k(x)_{terr}$ and $\theta(x)_{terr}$ as $\frac{1}{t-1}$ can be very small. These sudden changes can occur when there is a change in terrain type or weather conditions, but their impact will diminish with the passage of time. Manipulating Eq. (11.42), the recursive update rule can be obtained as follows:

$$\hat{\mathbf{X}}^{[t]} = P^{[t-1]} \sum_{i=0}^{t-1} \mathbf{Z}^{[i]}$$

$$= P^{[t-1]} \sum_{i=0}^{t-2} \mathbf{Z}^{[i]} + P^{[t-1]} \mathbf{Z}^{[t-1]}$$

$$= \frac{t-2}{t-1} \times \underbrace{\frac{1}{t-2} \sum_{i=0}^{t-2} \mathbf{Z}^{[i]}}_{\hat{\mathbf{X}}^{[t-1]}} + P^{[t-1]} \mathbf{Z}^{[t-1]} \tag{11.43}$$

$$= \hat{\mathbf{X}}^{[t-1]} + P^{[t-1]} \left(\mathbf{Z}^{[t-1]} - \hat{\mathbf{X}}^{[t-1]} \right).$$

Eq. (11.43) represents the recursive state update rule, wherein the term $P^{[t-1]} \left(\mathbf{Z}^{[t-1]} - \hat{\mathbf{X}}^{[t-1]} \right)$ represents the *measurement innovation*, i.e., the new information acquired via the new observation. Similarly, the recursive update rule for the filter response can also be derived as

$$P^{[t]} = P^{[t-1]} - P^{[t-1]} (P^{[t-1]} + 1)^{-1} P^{[t-1]}. \tag{11.44}$$

From Eqs. (11.43)–(11.44), it is clear that the filter is a modified moving average filter [21] with increasing window size, which accommodates all the data available. The predictions begin at $t = 2$, and $\hat{\mathbf{X}}^{[1]} = \mathbf{Z}^{[1]}$. Here, $\mathbf{X}^{[t]}$ is a function of $\mathbf{Z}^{[0:t-1]}$ which is a series of points indexed in time order, i.e., a time-series. So, a common time-series forecasting method can be used to estimate the value of $\mathbf{X}^{[t]}$ such as various variants of autoregressive moving average (ARMA) model [22]. In this case, a modified moving average model (ARMA(0,0,1)) was used that computes the average of all the data points available to estimate the value of $\mathbf{X}^{[t]}$.

4 Case 2: multi-rotor UAV operating in the presence of external disturbances

Similar to the case of UGVs, the term $k(t)_{env}$ in Eq. (11.38) is now replaced by its estimated value $\hat{k}(t)_{env}$ based on an autoregressive model presented in Eq. (11.43). Thus, the estimated maximum range for UAV for the remainder of the mission is now given by

$$\hat{d}_{max}^{[t:]} \triangleq \frac{\tilde{E}_{rem}}{\frac{P_{anc}}{v_{opt}} + \left\{\frac{\left[\frac{UAV\,\Omega_{man}\rho C_D A r_p^3}{6 k_r \sqrt{N_R}}\right] + \left[\frac{1}{r_p \sqrt{2 N_R \rho \pi}}\right]}{UAV\,\Omega_{man} v_{opt}}\right\} [m_{RG} + \hat{k}(t)_{env} m_{RG}]^{\frac{3}{2}}}.$$

(11.45)

Now, the maximum operational range estimation can be achieved similarly to Eqs. (11.41) and (11.43). However, in case of UAV, let $\hat{X}^{[t]} = \hat{k}(t)_{env}$ and $Z^{[t-1]} = k(t-1)_{env}$ define the estimated and observed values of the environmental variable, respectively, which are required for operational range estimation.

11.6 Experiments

In Sects. 11.5.4.1 and 11.5.4.2, *offline* and *online* range estimation models for the unified framework were presented. A precursor to the *offline* range estimation model was also discussed in Sect. 11.4.2. All three range estimation models were empirically evaluated in the following settings:

- **Simplified Range Estimation Model** was tested in indoor setting, which usually presented smooth surfaces $\theta \approx 0$ with constant average gradient of slope $\gamma \approx$ constant. The test-bed used for indoor experiments was *Rusti V1.0* as shown in Fig. 11.13.
- **Unified Framework** was designed to account for outdoor environment conditions along with unforeseen and unavoidable variations to missions that may have to be tackled in reality. Thus, this framework was tested in natural outdoor settings. The test-beds for outdoor experiments were *Rusti V2.0* shown in Fig. 11.14 and *ArDrone 2.0* shown in Fig. 11.15.

  - *Offline Range Estimation Model.* The parameters in Eqs. (11.23) and (11.23) were set based on a few prior experiments that had to be carried out independent of the actual field trials.

- *Online Range Estimation Model.* The parameters in Eqs. (11.23) and (11.23) were updated in real-time using the autoregressive moving average filter.

In what follows, firstly, the indoor experiments are presented for the simplified model, followed by outdoor experiments for the unified framework.

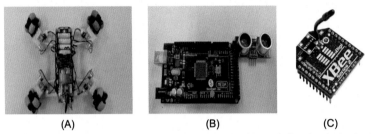

(A)   (B)   (C)

Figure 11.13 Rusti V1.0 with its sensory array: (A) Rusti V1.0 with omni-directional wheels; (B) ultrasonic sensor with Arduino Mega 2560 micro-controller; (C) short range wireless communication module (XBee) for indoor navigation. Image taken from [2].

(A)   (B)   (C)

Figure 11.14 Rusti V2.0 with all terrain wheels and an external 3-axis accelerometer sensor for outdoor field trials. Experiments were conducted on (A) asphalt, (B) grass and (C) tiles. Image taken from [16].

(A)   (B)

Figure 11.15 ArDrone 2.0 with GPS used for outdoor experiments. Experiments were conducted in (A) outdoor parking lot and (B) public park. Image taken from [16].

11.6.1 System identification

Before presenting the empirical analysis for indoor and outdoor field trials, a summary of system identification parameters is given in Table 11.1.

Table 11.1 System Efficiency Calibration for Rusti V1.0, Rusti V2.0, and ArDrone 2.0.

Robot	$\neg \eta_1$	$\neg \eta_2$	$\neg \eta_3$	$\neg \eta_4$	$\Omega = \Pi_{i=1}^{4} \eta_i$
Rusti V1.0	99.5%	94.2%	9.2%	99.9%	8.615%
Rusti V2.0	99.5%	**	**	99.9%	**
ArDrone V2.0	99.5%	27%		99.9%	26.84%

In Table 11.1, the items marked with ** were not directly observable/measurable. Such terms were instead accounted for by clustering the variables in the equations and considering multiple terms in unison. Additionally, for the drone, the average propulsion efficiency was jointly obtained from a closely related work [9]. Further details can be found in [17].

11.6.2 Indoor experiments

The indoor experiments were performed using Rusti V1.0 equipped with $HC-SR04$ Ultrasonic ranging module. Box-shaped trajectories were executed for planar surfaces and oscillating trajectories were executed for elevated surfaces as illustrated via Fig. 11.16. In order to validate the ancillary power consumption model given by P_{anc} in Eq. (11.6), the ultrasonic sensor was operated at various operational frequencies, and the power consumed was plotted as shown in Fig. 11.17. The model for ancillary power consumption was verified empirically. The power consumption model obtained is

$$P_{sensor} = 5.7318e^{-5} f_s + 0.0293. \tag{11.46}$$

From Eq. (11.46), it becomes clear that the power consumed by the sensor array in idling state is 0.0293 W. The worst case power consumption at a sampling rate of 100 Hz was found to be 0.0348 W. The power consumption of the micro-controller unit (MCU) which controls the ultrasonic ranging sensor and the wireless communication (XBee) was found to be quite stable at 0.3928 W with XBee consuming 0.166 W, irrespective of the size of data being transmitted. Thus, the overall ancillary power consumption model now becomes

$$P_{anc} = \underbrace{\{5.7318e^{-5} f_s + 0.0293\}}_{P_{sensor}} + \underbrace{\{\underbrace{0.166}_{P_{XBee}} + \underbrace{0.3928}_{P_{MCU}}\}}_{P_C}. \tag{11.47}$$

In Fig. 11.18, the energy utilized as the robot covers more ground is showcased. Localization was disregarded for these experiments so that the computational energy

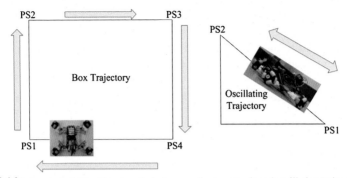

Figure 11.16 Indoor trajectories. Box-type trajectory on planar grounds and oscillating trajectory on evaluated plane; PS_i represents pit stops where the robot was made to stop for a pre-determined time period. This was done to emulate scenarios where a robot may need to stop and process data during a real mission.

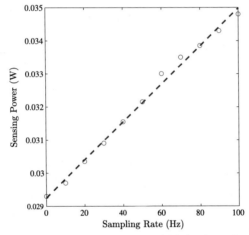

Figure 11.17 Energy consumed by sonar, illustrating modulation in ancillary power consumed by $HC-SR04$ Ultrasonic ranging module when operating at various frequencies f_s. Image taken from [2].

could be quantified. Quantifying the energy consumed by the robot for performing SLAM, localization, etc., is beyond the scope of this book, as it requires low level cache management for the on-board processors and also depends on the nature of the source code used. Besides, no matter how accurately these factors are accounted for, the energy that goes into these components does not contribute to the maneuvering energy, and hence, the author(s) chose to disregard them. An interesting fact to note in this figure is that, owing to lack of localization, the robot drifted from its assigned path especially at high duty cycles, which was even more pronounced on the evaluated plane. If the robot were to precisely follow the box trajectories and the oscillating trajectories, then the trends would be quite linear. Nonetheless, this was not the objective of the experiments. The trajectories were pre-coded since the robot was not meant to be fully autonomous, but in real-world scenarios a robot usually does not follow repetitive trajectories. Also, the distance covered is maximum when the duty cycle is

100%, whilst it decreases as the robot spends more time stopping for gathering and processing information. This is in accordance with the definition of the duty cycle.

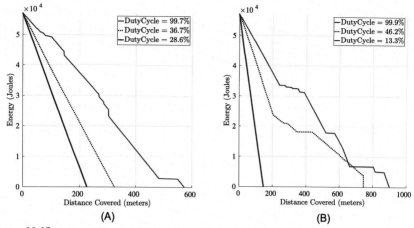

Figure 11.18 Energy utilization for (A) flat plane and (B) elevated plane experiments. Image taken from [2].

Fig. 11.19 shows the estimation error for the simplified range estimation model for both planar and elevated environments. In this figure, the negative values indicate underestimation, i.e., achieved range was larger than the estimated values, whilst the positive error represents the opposite situation. The accuracy of this model was evaluated to be 66%–91%. However, as the duty cycle is reduced further, the model sometimes has trouble to precisely estimate the power consumption for ancillary functions, which leads to erroneous estimates for the achievable range. The reason for this could be attributed to the fact that, at lower duty cycles, the robot transmits larger amounts of data, thus the ancillary power must also account for the data transmission rate, which is beyond the scope of the current model but was accounted for in the

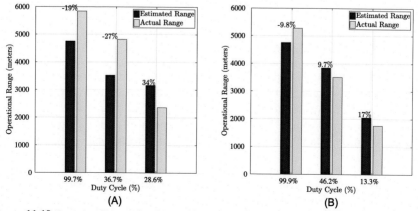

Figure 11.19 Operational rage estimation performance for (A) flat plane and (B) elevated slope experiments. Image taken from [2].

unified framework. Additionally, on-board vibrations introduce noise in sensor data being acquired, which also affects the model performance. Thus, the performance was logged for very noisy data.

The component-wise power consumption was also evaluated, for which a detailed breakdown is shown in Table 11.2. The maneuvering power was estimated for a fixed average speed of ≈ 1 m/s and the sensing power was estimated for sampling rates ranging from 0 to 100 Hz. For the case of 0 Hz, a low-power sleep mode was implemented for the sensor to significantly reduce the power consumption. This could come in very handy later on when designing a controller/real time scheduler that can divert the power from ancillary branch to maneuvering branch to further enhance the achievable range by cutting down unnecessary power consumption.

The minimum and maximum percentages of a component's power to the total system power are shown in Table 11.2. To find the minimum, the component is assumed to be running at its minimum power whilst the rest of the system consumes the maximum power, and the opposite is assumed for the maximum [23]. From this table, it can be seen that when on-board computation is not required, 90% of the total power is consumed by the motors for maneuvering. However, when adding an embedded computer to the ancillary branch (which consumes 8–15 W power [23]), this composition will go down significantly (30.5%–36.5%). This composition will be further affected by varying velocities for traversal and type of sensors. Thus, accurately estimating the range is all the more critical in such cases.

Table 11.2 Power consumption breakdown.

Component	Power (W)	Composition (%)
Maneuvering	4.8158–6.8456	89.13–92.02
Sensing (Ultrasonic Senor)	0.0293–0.0348	0.46–0.54
Wireless Communication	0.165–0.166	2.23–3.05
Micro-controller unit (MCU)	0.3928	5.29–7.26

11.6.3 Outdoor experiments

In this section, the author explains the outdoor experimental conditions in which the unified framework from Sect. 11.5.4 was evaluated. For the UGV, 36 experiments were carried out on various terrains, comprising grass, tiles, or asphalt with varying elevations and wheel rpm of 80 and 140.[8] The reason for considering these terrain types individually was the lack of capable hardware to determine change in terrain

[8] Given the wheel radius of 65 mm, these translate to $v = 0.544$ and 0.952 m/s, respectively. The velocities were pre-set at the beginning of the field trial and were not monitored during the field trial. For the *offline* and *online* models, the heading velocity remained constant, and for turning, while one side of motors were slowed by δ, the other side was sped up by the same factor. This ensured that the average velocity of the center of mass of the robot remained constant.

types on-the-fly and accordingly adjust the coefficient of rolling friction for making online prediction. This could be achieved if a camera and LIDAR were to be used to subtract the background information and match the features of the foreground with pre-selected images of the terrains that can be seen in real field trials. Such ideas have been explored in works like [24] that use LIDAR and camera to complement each other to detect obstacles and classify the terrain. However, this procedure fails in the absence of proper lighting conditions, and, additionally, estimating the rolling friction coefficient was beyond the scope of this work.

For system efficiency calibrations, 6 *minimal load* tests at 100 and 200 rpm were performed. The average rolling coefficients for terrain resistance offered by grass, tiles, and asphalt were set as 0.099, 0.066, and 0.062, respectively, and the prior information of γ to be used in the *offline* estimation was set based on Table 1 of [10]. In order to deduce the average values of the parameters, $\overline{k}_{terr}^{[0:t]} \leftarrow \dfrac{\tilde{E}^{[0:t]} - AE^{[0:t]}}{F_{slopes}}$. Similarly, the equations for $\overline{\gamma}^{[0:t]}$ and $\overline{\theta}^{[0:t]}$ can be deduced.

As for the UAV, 30 field trials were executed which were split into two different sets, viz., *hovering* and *motion*. For *hovering*, only the altitude of the UAV was varied and the human operator occasionally had to send control commands to maintain the position of the drone within a set perimeter. As opposed to this, in *motion* case, the human operator constantly fed linear motion commands to the drone whilst occasionally commanding the drone to hover (in cases when wind gusts lead to dangerously high velocity gains). This not only helped ensure the safety of the drone and its operator, but also helped emulate the real life scenarios in which the drone may loose connection to the base station (Fig. 11.9(B)) or corruption of mission critical information (Fig. 11.9(A)). The control commands were sent at operator-defined data transmission rate such that $P_{communication}$ in Eq. (11.15) remained constant. Ten experiments for *hovering* at different altitudes varying in-between [1, 10] meters were carried out. Furthermore, 20 tests at 5 different operational velocities[9] for *motion* to account for a mix of wind gusts, altitude adjustments, variable mission speeds, and trajectories were also considered. Experiments were performed at intervals of 2 hrs so as to account for changing environmental factors like wind and weather conditions.

In case of UAV, the wind compensation angle of the UAV was constantly monitored to evaluate the adjustments in the thrust that the robot needs to make to maintain flight stability. Through simple geometry, this was then used to calculated real-time values of $k(t)_{env}$ as explained in Fig. 11.20. From this figure, it can be seen that the net stabilization required on the part of the rotorcraft is $CB = OB(1 - \cos(\theta))$. When the rotorcraft was maintaining a constant altitude, $OB = m_R g$. So here CB represents $k(t)_{env} m_R g$ as explained in Eq. (11.27). Therefore, $k(t)_{env} = 1 - \cos(\theta)$.

---

[9] The translation velocities were chosen from [0.1, 0.2, 0.4, 0.6, 0.8] m/s which are subject to brief change upon change in heading direction. For instance, consider operational velocity of 0.1 m/s along $+X$ direction. Upon request to change direction to $-X$, the velocity during this brief transition period will vary from 0.1 m/s in $+X$ to 0 m/s in $+X$ followed by 0.1 m/s in $-X$ direction. This velocity profile cannot be feasibly estimated for the *offline* model, so, the operational velocity was used directly (0.1 m/s for this example), while for the *online* model, velocity was continuously observed, so, the average velocity till current time step was used.

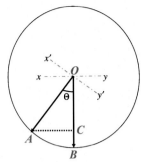

xy	*Initial orientation*
x'y'	*Adjusted orientation*
CB	*Net Adjustments*
OB	*Rotor thrust (T)*
OA	*Adjusted thrust*
θ	*Wind compensation angle*

Figure 11.20 Geometric analysis of wind compensation angle to deduce the value of parameter $k(t)_{env}$. Suppose the UAV is stable, then the orientation is represented by xy and OB represents the thrust (T) exerted by the UAV for maintaining flight. Now, assume that because of sudden wind action, the rotorcraft is displaced by an angle θ (which can also be interpreted as wind compensation angle) and the new orientation is $x'y'$. So, OA will represent the same thrust under the sudden influence of the wind at an angle θ to the previous direction. Thus, the net altitude destabilization effect of the wind is given by BC. Image taken from [16].

For Eqs. (11.23) and (11.38), as was discussed earlier, to make predictions of d_{max}, the average resistive forces need to be approximated, which in turn are factors of $k(x)_{terr}$ and $k(t)_{env}$. Therefore, to estimate these parameters in real-time, the ARMA filter is fed with real-time mission data.

11.6.4 Batteries used for field experiments

Since the *ArDrone* comes factory fitted with a mini-tamiya connector, the stock battery, i.e., 11.1 V @ 1500 mAh high density LiPo battery, was used for it. However, having custom built *Rusti V2.0*, the following two LiPo batteries were considered for field trials:

- 11.1 V @ 2200 mAh
- 11.1 V @ 1500 mAh (also used for *ArDrone*)

Subsequent sections discuss the results obtained for the UGV followed by UAV during the outdoor field trials.

11.6.5 Case 1: UGV

In Figs. 11.21–11.23, both the *offline* and *online* models are pegged against the true achieved range during real field trials on grass, asphalt, and tiles, respectively. For *offline* estimation, based on Eq. (11.23), it was previously explained that for estimating the operational range, the model needs prior information about θ, γ, and k_{terr}. Also, the mechanical efficiency[10] is unascertained. The value for γ was obtained based on Table 1 of [10], and that of θ_{avg} was empirically set to 5°. Estimating the values of

---

[10] *Rusti* Ω_{man} is a factor of only motors and will account for both frictional losses, as well as heat losses in motors.

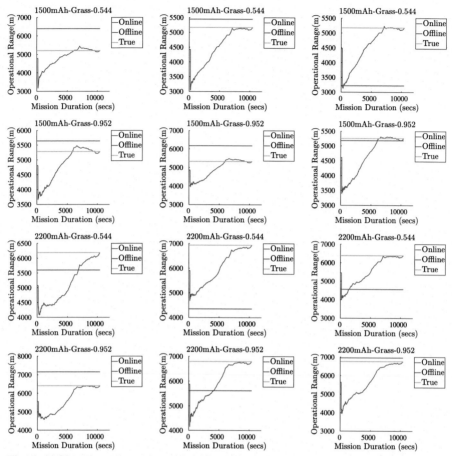

Figure 11.21 Rusti's operational range estimation for grass. Rows 1–2 represent experiments performed using 11.1 V @ 1500 mAh battery @ 80 rpm followed by @ 140 rpm. Rows 3–4 represent a similar pattern for 11.1 V @ 2200 mAh battery, respectively. Image taken from [16].

k_{terr} and $^{Rusti}\Omega_{man} = \Pi^3_{i=2}\neg\eta_i$ is rather challenging and requires some prior field experience. The author(s) chose to modify Eq. (11.23) such that the maneuvering efficiency term, i.e., $^{Rusti}\Omega_{man}$, is now considered within the integral, and the new term $\frac{k_{terr}}{^{Rusti}\Omega_{man}}$ was treated a single terrain-dependent variable. The average value of this terrain dependent factor was then determined through a series of field trials as $\frac{k_{terr}}{^{Rusti}\Omega_{man}} = [3.09, 2.81, 2.69]$ for grass, asphalt, and tiles, respectively. For *online* estimation, the belief of the model over the net operational range achievable is updated in real-time based on cumulative performance characteristics. Effectively, on average, the net true distances covered by the robot at 0.544 m/s and 0.952 m/s are almost the same on all types of terrain. This can be attributed to the fact that, because of the use of high torque DC motors in *Rusti V2.0*, the net ancillary energy requirements are negligible compared to maneuvering energy (which is independent of operational speed). From Eq. (11.23), it can be seen that this, in fact, will be the case if $P_{anc} \ll P_{man}$.

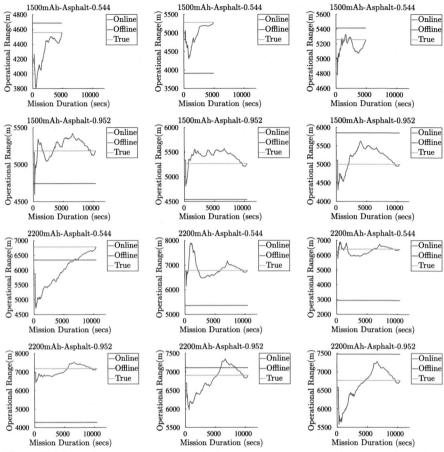

Figure 11.22 Rusti's operational range estimation for asphalt. Rows 1–2 represent experiments performed using 11.1 V @ 1500 mAh battery @ 80 rpm followed by @ 140 rpm. Rows 3–4 represent a similar pattern for 11.1 V @ 2200 mAh battery, respectively. Image taken from [16].

Also, the true distance covered using the 2200 mAh battery is greater than that covered using the 1500 mAh battery, however, they are not in proportion of the battery capacities, i.e., the ratio of battery capacities is $\frac{22}{15} = 1.47$ but the ratio of achieved true range is $\frac{6.67}{5.17} = 1.29$. This difference can be attributed to the fact that the mass of the robot is slightly higher when using the 2200 mAh battery which dilutes the effect of extra charge capacity. However, these proportions will be drastically affected so much so that the maneuvering energy consumption could account for less than 50% of the net energy supplied by the battery [23]. Having said this, it must also be pointed out here that this proportion depends on the mission profile and the load borne by the sensor array being used. These are beyond the scope of this work as the robot with minimal sensor array was used for the analysis of operational range. Nonetheless, if the proportions were to change as mentioned earlier, then the need for high accuracy in operational range estimation becomes ever more pronounced.

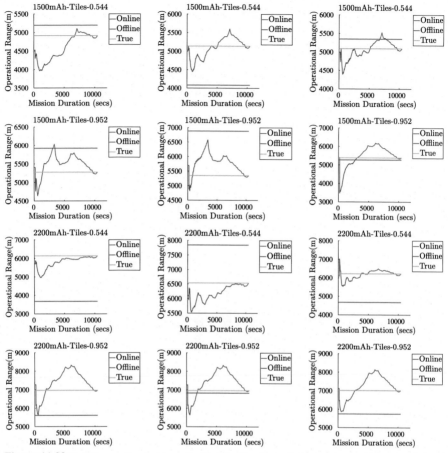

Figure 11.23 Rusti's operational range estimation for tiles. Rows 1–2 represent experiments performed using 11.1 V @ 1500 mAh battery @ 80 rpm followed by @ 140 rpm. Rows 3–4 represent a similar pattern for 11.1 V @ 2200 mAh battery, respectively. Image taken from [16].

Then in Fig. 11.24, a bar plot is shown to evaluate the average accuracy of both the proposed *offline* and *online* models, along with their respective variances. As expected, the *offline* model tends to over- or under-shoot the true operational range incurring extreme errors with high variance, whilst the *online* models tend to attain the true operational data with a very high accuracy and low variance. It must be pointed out here that, while traversing on grass using the following settings: 1500 mAh @ 0.952 m/s, both models show comparable average performance, while for 2200 mAh @ 0.952 m/s, the *offline* model performs slightly better. Despite this, the variance of the *offline* model remains higher which can also be confirmed from Fig. 11.21. Overall, the *online* model shows ≈ 60% enhanced accuracy as compared to its adversary for operational range estimation of Rusti V2.0.

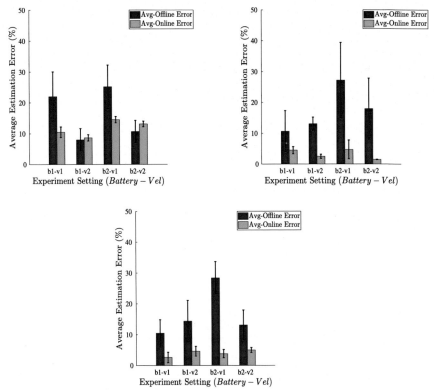

Figure 11.24 Range estimation error for Rusti while traversing on grass, asphalt, and tiles, respectively. Here $b1$ and $b2$ refer to 1500 and 2200 mAh batteries, while $v1$ and $v2$ refer to 0.544 and 0.952 m/s velocities, respectively. Image taken from [16].

11.6.6   Case 2: UAV

Fig. 11.25 demonstrates the real world performance of the *offline* and *online* estimation models for the case of a UAV.

For *offline* estimation, prior information regarding the operational environment of the robot is required to make meaningful predictions of its operational range. Since there is no prior research explaining how the values of the parameter $k(t)_{env}$ vary, an additional set of 5 experiments (each in varying conditions) were performed and their data was averaged for estimating the value of the parameter $k(t)_{env}$ which was found to be ≈ 0.01. Using this prior information, the maximum operational range was obtained using the *offline* model for the 20 experiments presented here. As the mean estimation error for the *offline* model is about 30 meters in each experiment, with even lower errors at slower speeds, the value of $k(t)_{env} = 0.01$ is claimed to be optimal.

For *online* estimation, the estimate of the net operational range is updated using the autoregressive average of the data acquired in real time. Thus, no prior mission information was deemed necessary to deduce the value of the parameters $k(t)_{env}$ and v. Instead, they are deduced based on real-time mission information. Upon take-off, the

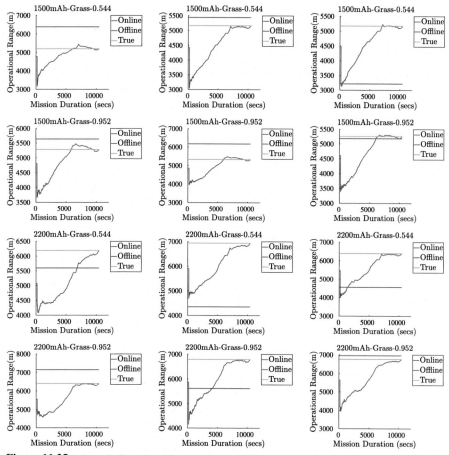

Figure 11.25 ArDrone's Operational Range Estimation. First row represents experiments carried for $v =$ 0.1, followed by $v = 0.2, 0.4, 0.6$, and 0.8 m/s, respectively. Columns represent multiple experiments at corresponding v. **Note that the results are only comparable across columns, owing to different scales of plots across rows.** Image taken from [16].

drone is initially quite unstable owing to significantly high rotor rpm required for lift-off which then settles to a stable rpm for hovering. During this time, not only the in-house electronics of the drone are unreliable, but also the GPS sensors used need time to sync with satellite information. This initial instability in the ArDrone and the sensor's data just after take-off is what the author calls as the *burn-in* phase. The data of the burn-in phase is discarded and the *online* estimation framework is activated only upon ArDrone's stabilization. For the purpose of continuous representation in graphs, the data was interpolated during this phase, resulting in the initial straight line trends observed during the burn-in phase of the plots. Also, as is evident from the plots, the *online* estimator converges to the true distance as the mission progresses. As more and more mission data becomes available, the estimation performance of the *online* model becomes significantly better than that of its *offline* counterpart.

It must also be pointed out here that the variations in the *online* estimator are quite profound during the early stages of the mission which can be attributed to the fact that the estimator is trying to update its belief with sparse and limited amount of data, but it quickly stabilizes as the amount of data grows. Also, it might seem that increasing the operational velocity (v) always leads to an increase in the operational range (d). However, when the ArDrone attains an operational velocity v_{opt} which is high enough, so much so that the aerodynamic drag forces acting on the body of the drone are higher than those on the propeller, the theoretical maximum operational range (d_{max}) will be attained and any further increase in the velocity will result in a decrease in the operational range. Besides, such high velocities (v_{opt}) are not attainable by current multi-rotor UAVs.

The authors would also like to highlight that the variance in the input data (wind compensation angle), being very low, results in low variance of $k(t)_{env}$, which, coupled with its low absolute value (1% on average), results in very low variance of predicted distance, often less than a meter. Therefore, for the sake of clear understanding of the readers and legible visual representation of the field trials, the variance in predicted distance was omitted.

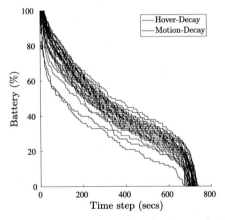

Figure 11.26 Battery decay for UAV while hovering and motion. Image taken from [16].

Additionally, Fig. 11.26 showcases how the energy stored in the battery is consumed as the mission progresses. An interesting fact to note here is that the trends for both the *hovering* and *motion* cases are quite similar. The reason for this can be attributed to the fact that the value of $k(t)_{env}$ which represents the average excess percentage of thrust that needs to be exerted to maintain stability and velocity, owing to changing environmental conditions, remains below 2%.[11] Therefore, the major component of maneuvering energy is utilized to maintain flight, instead of stabilizing the rotorcraft and maintaining its velocity.

Finally, Fig. 11.27 shows bar plots for the operational range estimation performance for both the discussed *online* and *offline* models. For this, the average estima-

---

[11] Analyzed from the data obtained from *motion* experiments at varying velocities as shown in Fig. 11.25.

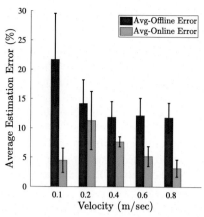

Figure 11.27 Range estimation error. Plot showing error in operational range calculated using the *offline* and *online* models along with corresponding standard deviation. Image taken from [16].

tion error of both frameworks for each operational velocity is shown. It can be clearly seen from the graph that the *online* model is $\approx 58\%$ more efficient than its counterpart. Also, to clarify the high *offline* estimation error at $v = 0.1$ m/s, the author(s) would like to point out that, even for small amounts of *hovering* time, the percentage difference in the average velocity and operational velocity is considerably higher than for higher velocities, which translates to higher percentage error in prediction through the use of the *offline* model.

11.7 Summary

This chapter addressed one very critical aspect of autonomous navigation for mobile robots. This will help prevent the mobile robots from failing to complete the task and being strangulated amidst the field. The problem being addressed can be summed up under the keyword "operational range estimation" for which the authors discussed two frameworks:

- **Simplified Framework.** This framework was designed explicitly for ground robots operating on smooth terrains with fixed gradient, which is usually the case for indoor environments. This framework comprises two components, viz.,
  - *Simplified Energy Distribution Model*, which explains how the energy is distributed throughout the robot and all its components. The maneuvering energy model accounts for planar and elevated terrains, while the ancillary energy model includes energy consumed by sensors along with the unwarranted losses. This model can be used to deduce the net energy available for traversal.
  - *Simplified (Offline) Range Estimation Model*, which transforms the net traversal energy into operational range and also proposes a theoretical upper bound for it.
- **Generic (Unified) Framework.** This framework was presented as a further enhancement over the simplified variant and encompassed a variety of robots operat-

ing in a myriad of environmental conditions (harsh and otherwise). This framework generalizes the models of the simplified models as follows:

- *Generic Energy Distribution Model* extends the previous variant of maneuvering energy model to various classes of robots. Additionally, the ancillary energy model now accounts for data transmission rate for short range wireless communications.
- *Generic Range Estimation Model,* as opposed to the previous *offline* model for smooth terrains, this model now has an *offline* variant capable of handling uneven terrains and unforeseen environmental disturbances. Not only this, an *online* model was also proposed to account for sudden changes in the mission profile as they present themselves. Both the extensions were studied in depth for UGVs and UAVs.

The strengths of the **Generic (Unified) Framework** are highlighted below:

- The unified framework is equally applicable to both commercial and custom-built robots alike, provided additional sensors can be incorporated to log the appropriate data.
- The concept of duty cycle proposed herewith brings this model really close to real-world scenarios, making the framework applicable without hassles.
- Having obtained the average accuracy of almost 93.87% with the online variant and 82.97% with the offline variant, it is safe to conclude that the framework is by far the state-of-the-art operational range estimation framework for all robots that may be considered for field trials.

All what remains now is to couple this framework with energy efficient path planners and then the robots can be guaranteed to return to the base station by the end of their mission (not accounting for impromptu hardware failures).

References

[1] https://commons.wikimedia.org/wiki/File:Military_Working_Dog_Aggression_Training_ 161104-M-QX129-048.jpg.

[2] Kshitij Tiwari, Xuesu Xiao, Chong, Nak Young, Estimating achievable range of ground robots operating on single battery discharge for operational efficacy amelioration, in: 2018 IEEE/RSJ International Conference on Intelligent Robots and Systems (IROS), IEEE, 2018, pp. 3991–3998.

[3] K. Tiwari, X. Xiao, V. Kyrki, N.Y. Chong, ORangE: Operational Range Estimation for mobile robot exploration on a single discharge cycle, in: Robots in the Wild Workshop: Challenges in Deploying Robust Autonomy for Robotic Exploration, Robotics: Science and Systems (R:SS), 2019.

[4] D. Panigrahi, C. Chiasserini, S. Dey, R. Rao, A. Raghunathan, K. Lahiri, et al., Battery life estimation of mobile embedded systems, in: VLSI Design, 2001. Fourteenth International Conference on, IEEE, 2001, pp. 57–63.

[5] F. Zhang, G. Liu, L. Fang, Battery state estimation using unscented Kalman filter, in: Robotics and Automation, 2009. ICRA'09. IEEE International Conference on, IEEE, 2009, pp. 1863–1868.

[6] M.H. Chang, H.P. Huang, S.W. Chang, A new state of charge estimation method for LiFePO4 battery packs used in robots, Energies 6 (4) (2013) 2007–2030.

[7] Q. Miao, L. Xie, H. Cui, W. Liang, M. Pecht, Remaining useful life prediction of lithium-ion battery with unscented particle filter technique, Microelectronics and Reliability 53 (6) (2013) 805–810.

[8] L. Liao, F. Köttig, Review of hybrid prognostics approaches for remaining useful life prediction of engineered systems, and an application to battery life prediction, IEEE Transactions on Reliability 63 (1) (2014) 191–207.

[9] A. Abdilla, A. Richards, S. Burrow, Endurance optimisation of battery-powered rotorcraft, in: Conference Towards Autonomous Robotic Systems, Springer, 2015, pp. 1–12.

[10] A. Sadrpour, J. Jin, A.G. Ulsoy, Mission energy prediction for unmanned ground vehicles using real-time measurements and prior knowledge, Journal of Field Robotics 30 (3) (2013) 399–414.

[11] K. Tiwari, X. Xiao, N.Y. Chong, Estimating achievable range of ground robots operating on single battery discharge for operational efficacy amelioration, in: 2018 IEEE/RSJ International Conference on Intelligent Robots and Systems (IROS), IEEE, 2018, pp. 3991–3998.

[12] Y. Mei, Y.H. Lu, Y.C. Hu, C.G. Lee, Energy-efficient motion planning for mobile robots, in: Robotics and Automation, 2004. Proceedings. ICRA'04. 2004 IEEE International Conference on, vol. 5, IEEE, 2004, pp. 4344–4349.

[13] D. Brooks, V. Tiwari, M. Martonosi, Wattch: A Framework for Architectural-Level Power Analysis and Optimizations, vol. 28, ACM, 2000.

[14] O. Tremblay, L.A. Dessaint, A.I. Dekkiche, A generic battery model for the dynamic simulation of hybrid electric vehicles, in: Vehicle Power and Propulsion Conference, 2007. VPPC 2007, IEEE, 2007, pp. 284–289.

[15] R. Mur-Artal, J.M.M. Montiel, J.D. Tardós, ORB-SLAM: a versatile and accurate monocular SLAM system, IEEE Transactions on Robotics 31 (5) (2015) 1147–1163, https://doi.org/10.1109/TRO.2015.2463671.

[16] K. Tiwari, X. Xiao, A. Malik, N.Y. Chong, A unified framework for operational range estimation of mobile robots operating on a single discharge to avoid complete immobilization, Mechatronics 57 (2019) 173–187.

[17] K. Tiwari, X. Xiao, A. Malik, N.Y. Chong, A unified framework for operational range estimation of mobile robots operating on a single discharge to avoid complete immobilization, Mechatronics 57 (2019) 173–187.

[18] S.W. Kim, Y.H. Lee, Combined rate and power adaptation in DS/CDMA communications over Nakagami fading channels, IEEE Transactions on Communications 48 (1) (2000) 162–168.

[19] M. Gatti, F. Giulietti, M. Turci, Maximum endurance for battery-powered rotary-wing aircraft, Aerospace Science and Technology 45 (2015) 174–179.

[20] A. Abdilla, A. Richards, S. Burrow, Power and endurance modelling of battery-powered rotorcraft, in: 2015 IEEE/RSJ International Conference on Intelligent Robots and Systems (IROS), 2015, pp. 675–680.

[21] H. Sato, Moving average filter, 2001. US Patent 6,304,133.

[22] D. Graupe, A.A. Beex, G.D. Causey, ARMA filter and method for designing the same, 1980. US Patent 4,188,667.

[23] Y. Mei, Y.H. Lu, Y.C. Hu, C.G. Lee, A case study of mobile robot's energy consumption and conservation techniques, in: Advanced Robotics, 2005. ICAR'05. Proceedings. 12th International Conference on, IEEE, 2005, pp. 492–497.

[24] R. Manduchi, A. Castano, A. Talukder, L. Matthies, Obstacle detection and terrain classification for autonomous off-road navigation, Autonomous Robots 18 (1) (2005) 81–102.

Part IV

Scaling to multiple robots
Multiple robots for efficient coverage

Contents

This part goes a step further into the multi-robot setup and details the challenges that arise when scaling up to several robots to efficiently cover the target area.

IV.1 Multi-robot systems

Whilst scalability assists with the efficient area coverage, it comes at a cost. The cost is not just monetary (for purchasing the robots), but also refers to the additional efforts required for the team management. This chapter encompasses such challenges and describes some of the team management strategies that are widely used in the wireless sensor networking domain, but their connotations have been adapted to suit the multi-robot environment exploration setting. Additionally, a complex combination of such strategies, i.e., a mixture of multiple such strategies, is also discussed, which might come in handy for some end-user applications, and one such strategy is, in fact, being used in the scope of this book as well.

IV.2 Fusion of information from multiple robots

Multiple robots, each behaving as an expert, eventually lead to multiple learnt models which need to be either merged or somehow the best one needs to be selected, as all of them were essentially monitoring the same environment. To this end, this chapter builds upon the *information never hurts* principle and discusses how multiple such models can be effectively fused to have one globally consistent model at the end.

Multi-robot systems
The cost of scalability

12

Everything in life has a prize tag attached to it. So, the more the merrier, but, at what cost?

Dr. Kshitij Tiwari

Contents

Highlights

- Advantages of scaling to multiple robots
- Challenges when scaling up the team size
- Technical terms for defining the nature of co-ordination (if any) amongst the team
- Various communication protocols amongst peers, if communication is desired
- Tackling rogue agents to ensure optimal team performance

While *the more the merrier* is a useful policy to increase the efficiency in an *Intelligent Environment Monitoring (IEM)* setting, by deploying multiple robots to span a wider portion of the target area, it comes at a cost of its own. The cost, pertaining to monetary and practical challenges, is described in detail below. Thus, when designing the optimal deployment strategy using a scaled up robot fleet consisting of multiple units, these costs must also be weighed carefully.

12.1   Advantages of scaling

There are several advantages when scaling up the size of the robot team. They are discussed below.

Multi-Robot Exploration for Environmental Monitoring. https://doi.org/10.1016/B978-0-12-817607-8.00027-7

- **Distributed computing.** While using a single robot would require either storing and processing all data acquired on-board or transmitting the data to a base station which would then process the data at an additional communication overhead, using multiple robots distributes this computation. Each robot is only expected and capable of acquiring and processing a subset of the available data, thereby easing the computational burden per member of the team. In doing so, a variety of methods can be devised to assimilate the distributed information into a single globally consistent structure.
- **Wider area coverage.** Having multiple robots also allows for spanning a wider area. Each robot inherently has limited area coverage, owing to its resource constraints like flight time, battery, payload, etc. With multiple such resource constrained robots, the possibility to observe a wider portion of the target phenomenon is made available.
- **Temporally optimal coverage.** Temporal constraints, be it imposed from a time-critical mission or the resources that limit the maximal amount of information that can be accrued during the mission. This is rather challenging for a single robot, but with multiple robots, each robot being locally (temporally) optimal contributes to the overall temporal optimality of the team which, in theory, should be better than a single robot.
- **Heterogeneous coverage.** Given the different strengths of different mobile robot platforms, scaling up the team to incorporate multiple robots also provides for a chance to harness the unique strengths of each of these robot platforms. For instance, the agility of an aerial robot could be coupled with the payload capacity of a ground robot which in unison provide a diverse functional team as opposed to a single robot.

12.2 Challenges to scaling

Scaling up the size of the robot team has both pros and cons. The benefits stem from the increased agility and coverage capability of the team as the size of team grows. The downside is the monetary cost of setting up the team and selecting an apt deployment strategy. Deployment encompasses both communication and self-organization strategies, along with the physical deployment of the team in the field to gather observations. In what follows, some of the well-established control and communication strategies are described.

12.2.1 Selecting optimal team control strategy

Depending on where the decision making authority rests in the team, there could be multiple team management strategies that can be considered. Most of them are heavily used in the paradigm of mobile wireless sensor networks and information theory domains, and can be straightforwardly adapted to suit the context of this book. Below, such strategies are explained with the help of illustrative examples to ease the reader's understanding.

12.2.1.1   Centralized strategies

A typical centralized system has all the decision making power at one specific node and looks like Fig. 12.1. Usually there is one *master* node acquiring all the information from multiple *slave* nodes. The job of *slave* nodes is to passively gather the data and simply pass on to the *master* where all the processing and decision making happens. In this example, the CEO can be thought of as the *master* node and the employees can be thought of as *slave* nodes. As is the case in most small-scale startups, there is a workforce comprising the business owner and a handful of employees who directly report to the owner. As the stakes are very high, most of the decision making power rests with the owner. In fact, this is what is formally referred to as the chief executive/estimation officer (CEO) problem [1] in the information theory literature. An extension to the original problem can be found in [2] which considers secrecy constraints wherein the information transmitted by some employees gets intercepted by an unknown adversary (like a corporate spy) who eavesdrops on their correspondences with the CEO to learn as much about the business model as possible.

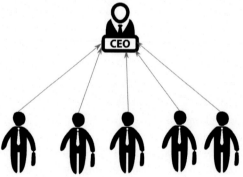

Figure 12.1 Illustration of a centralized topology via a small-scale startup structure. At the center is the CEO and at the bottom are multiple employees that directly report to the CEO.

This setting has several pros (+) and cons (−) like:

+   Very simple architecture and easy to maintain
+   Easy to set up and replicate
−   Single point of failure, e.g., if the CEO is sick or away for a long vacation, the entire business comes to a halt until the CEO resumes charge
−   Not scalable, e.g., if the startup were to have 100 employees and maintain the same architecture, the CEO will be stretched thin and will definitely not have time to address all employees
−   Susceptible to eavesdropping as explained by the corporate espionage example above

12.2.1.2   Decentralized strategies

Some of the limitations such as a single-point of failure and scalability can be ad-dressed by modifying the connectivity of the *master* to the *slaves*. For this, consider an additional set of nodes called the *manager* nodes which have the capability to assimilate some of the information locally and make decisions based on them. The global decision is eventually made by the *master* node but the core of such a structure is the delegation of decision making authority. This is best explained by the illustration in Fig. 12.2. In this setting, a medium-scale corporate business is considered. Herein, there are three levels of the hierarchy: (i) CEO, (ii) managers, and (iii) employees. Such a setup is known as decentralized.

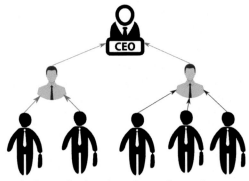

Figure 12.2 Illustration of a decentralized topology via a medium-scale corporate structure. At the top is the CEO, in the middle are managers who manage local teams, and at the bottom are regular employees.

While this setting overcomes some of the challenges of the *centralized* setup, this setup has its own pros (+) and cons (−) like:

+   Multiple decision making points
+   No single point of complete failure
+   Multiple local behavior gets aggregated to a global behavior
+   Compartmentalization of information: no single node with entire system informa-tion
+   Stable and fault-tolerant, e.g., if the CEO were to be away for a business trip, decisions can still be made and the company business will not come to a complete stand-still
+   Moderately scalable
−   Risk of duplication. As there is no co-ordinated decision making going on cen-trally, multiple nodes may generate similar/overlapping decisions. For instance, consider the manager on the left in Fig. 12.2 to be leading an IT team while that on the right to be leading an HR team. Both managers and their teams come up with an idea to develop an attendance system and are prone to duplication owing to lack of communication across teams.
−   Conflict resolution may become a challenge. Owing to decision making powers allocated to multiple managers, local managers from different departments who

have the same-level ranks may clash, and because they know they have independent decision-making roles, resolving inter-department conflicts may become more difficult. Aside from that, the same-level leaders who are competing may refuse to co-ordinate and cooperate with each other.

12.2.1.3 Distributed strategies

The distributed architecture can be thought of as a *divide-and-conquer* approach, wherein multiple nodes work collectively to achieve a common objective, however, locally they could be doing drastically different tasks. For instance, consider the multi-national company (MNC) setup as shown in Fig. 12.3. Here, an MNC called the XYZ Group has global head-quarters located in the center with multiple ancillaries spread across the globe. While each of the ancillary may be producing something completely different, e.g., one ancillary could be handling rubber and making tyres and the others could be handling pharmaceutical chemicals or manufacturing dairy products, all ancillaries fall under the global umbrella of XYZ Group and decision/production of each ancillary is reflected at the global HQ and peer ancillaries.

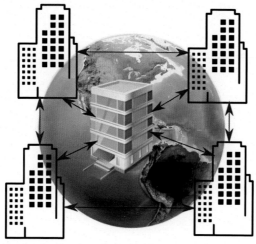

Figure 12.3 Illustration of a distributed topology via a large-scale multi-national corporation structure. Consider global head-quarters (HQ) located at the center with multiple ancillaries spread across the globe.

This setup has the following pros and cons:

+   Extensively scalable
+   Fault tolerant and robust to failures of local clusters
+   Computationally efficient
−   Difficulty in troubleshooting and deployment
−   Preliminary costs are high

12.2.1.4   Solitary confinement strategies

In all the previous descriptions, there was some level of control at different levels of the hierarchy. However, a considerably control-free setting is also possible, which, in fact, may be desired in some harsh communication-devoid operational conditions like sub-surface explorations, or in some conditions where the communication channels are unreliable and sporadic. This section defines such a setting.

Just as the name suggests, under the solitary confinement strategy, each robot is confined to a region within the target environment. In doing so, the human supervisor artificially manages the team by restraining the access rights of each robot. Each peer being limited to its own region, gets rid of the necessity to solve the obstacle avoidance challenges and global team co-ordination altogether. One potential approach to achieve this is to rely on Voronoi tessellation. As shown in Fig. 12.4, this would yield a fully tessellated target area where each robot is confined within its own Voronoi cell. As a further enhancement to this approach, in [3], the authors presented an adaptive variant wherein the Voronoi cells shrink/expand depending on hardware failures detected amongst peers. This ensures that the whole region is being consistently surveilled and falls under the category of *persistent environment monitoring (PEM)*.

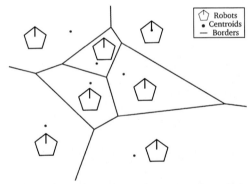

Figure 12.4 Illustrating solitary confinement strategy using Voronoi tessellation. Each Voronoi cell is patrolled by a single robot. The centroids of Voronoi cells are shown in red circle ●. Image based on [4].

The pros and cons of such a setting include:

+   Highly scalable
+   Maximal autonomy per peer
+   Confinement regions can be adapted dynamically (if needed)
−   Initial confinement regions needs to be pre-meditated
−   The smaller the region, the higher the redundancy in observations

12.2.2   Selecting optimal team communication strategy

The previous section primarily focused on the team management and self-organization strategies or, in other words, the control laws that could be adopted to suit robotic teams. The aim of this section then is to describe some of the team communication strategies which can be used when passing around information in the respective topologies described earlier. There are two aspects to consider while designing such communication strategies: (i) *What information will the agents pass around?* and (ii) *How will the information be passed around?* The nature of information, i.e., the answer to the former question is specific to the application being considered, but the latter can be addressed if an abstract notion of information is considered for the sake of this discussion. In the following sections, several such communication protocols are described.

12.2.2.1   Synchronous communication

Synchronous communication happens when the exchange of information is time co-ordinated. This is illustrated via Fig. 12.5. Here two users are communicating via a telephone channel which means that both of them must be using the respective handsets at the same time otherwise information exchange cannot progress.

Figure 12.5 Illustration of a synchronized communication strategy via a daily teleconversation scenario.

The pros and cons of such a communication protocol are:

+ Information exchange happens in real time with almost negligible delays
+ Response, if expected, is immediate
+ High throughput can be achieved
+ Minimal overhead as the crucial data/information can be directly transmitted
− Complex system is required, owing to apt synchronization requirements
− System complexity grows with the number of peers involved
− Does not scale well with the size of the team

As for the robotic applications of such a communication protocol, works like [5] have attempted to formalize the notion of synchronous communication for a set of locally connected robots that follow the *agree-and-pursue* protocol. Under this setting, robots are assumed to move on a circular ring, where the direction of motion is mutually agreed between locally connected agents. In further works, e.g., [6], the authors then analyzed the time complexity and deployment ability of some well-known co-ordination algorithms running on synchronous robotic networks like the move-toward-average, circumcenter, and centroid laws.

12.2.2.2   Asynchronous communication

Asynchronous communication as the name suggests progresses between transmitter(s) and receiver(s) without any co-ordination. This is illustrated in Fig. 12.6. Here the person on the left sends an email early in the morning, but the receiver only replies later at night. This is usually the case when people collaborate from distant locations spread across the globe.

Figure 12.6 Illustration of an asynchronous communication strategy via an email conversation scenario.

The pros and cons of this communication protocol are:
+ Information is transmitted when opportune to the sender
+ Easily scalable
− Prone to infinite waiting if there is no time-out set *a priori*, e.g., the email sent in the figure above lands in the spam box of the receiver, and the sender will wait until the receiver realizes this. This could take forever to get a response.
− Larger communication overhead. As opposed to the synchronous protocol, the data transmitted must also be appended/prefixed with flags used for allocating and controlling communication channels.

This communication protocol was demonstrated in [7], which uses a cloud-supported architecture for persistent surveillance based on multi-robots communicating and exchanging data asynchronously with the cloud. In [8], an asynchronous variant of space-time block code (STBC) [9] is discussed, which addresses the transmission energy efficient and reliability of cooperative robotic/wireless networks in an asynchronous manner. Similar architectures were considered in [10,11] which address the multi-agent rendezvous problem in a setting where robots take asynchronous action locally, yet are able to rendezvous in finite time.

12.2.2.3 Disconnected strategies

While in both the previous communication strategies there was some communication overhead incurred, in harsh communication devoid scenarios, not only the team co-ordination but also the intra-team communication is a challenge. To this end, this section describes a solution which is best suited for harsh conditions like sub-terrain explorations where communication channels are unreliable or, at times, even non-existent.

The disconnected strategy is perhaps the most lightweight communication protocol as it involves no communication across peers at all. This is illustrated via Fig. 12.7. With Tesla setting off the space-race, multiple companies are designing the space rockets to be the first to transport humans to outer space at affordable costs. As several million dollars worth of investment goes into the design and manufacturing of such space shuttles, multiple companies from start-up to corporate would like to keep their designs top-secret as long as they are in the race. For instance, two such designs from

Figure 12.7 Illustration of a disconnected management strategy using the space-race scenario.

different firms are shown in this figure, and the firms would have strict protocols that enforce them not to let information leak out or establish any sort of communication outside the firm, i.e., all information remains protected.

The pros and cons of this protocol are:

+ No communication overhead: either inter-peer or peer-to-base
+ No restriction to always maintain connectivity with peers or base
− High risk of redundancy
− Computational resources are local and limited

12.2.3 Tackling rogue agents

Aside from the monetary, team management, and team communication strategy considerations, another subtle yet important concern is that of *rogue* agents. Within multi-agent teams, some agents might go rogue during operation in the sense that their motors may be worn out amidst mission, sensors may be broken, or the likes as shown in Fig. 12.8. In the scope of this book, such agents are said to have gone rouge, and this will affect the environment monitoring capabilities of the team depending on the communication and control strategies being used. In [12], this problem was considered in a persistent monitoring setting wherein every part of the target environment needed to be monitored at all times, e.g., surveillance of international borders.

In such a setting, the authors have described the situation of how, if a team member goes rogue, the peers must compensate for that rogue agent.

Figure 12.8 Illustration of an team of (rogue) agents. Consider a 5-robot team tasked with a certain mission. During the mission, the robots on the bottom right and top left have damaged wheels/broken motors powering the wheels, and the robot in the center has a faulty sonar (labeled by the warning signs). These 3 robots are then said to have gone rogue.

12.3 Summary

This chapter addressed some practical issues that arise when scaling the robot team. Some of the important issues considered encompass the monetary cost of scaling up the team, team management, communication strategies, and handling of the rogue agents. Having said this, it is important to emphasize that not all of the aforementioned problems come with a solution, and hence some problem fall outside the scope of this book. Nonetheless, there are also advantages of scaling to multi-robot teams which motivated the upcoming chapter of this book. Within the scope of this book, the robot team is considered fully decentralized (computationally) and disconnected (no peer-to-peer or peer-to-base communication). This allows for the agents to explore the target area free of any communication constraints and pick the optimal observations that are best suited for the corresponding expert (robot) irrespective of others in the team.

References

[1] T. Berger, Z. Zhang, H. Viswanathan, The CEO problem [multiterminal source coding], IEEE Transactions on Information Theory 42 (3) (1996) 887–902.

[2] F. Naghibi, S. Salimi, M. Skoglund, The CEO problem with secrecy constraints, in: Information Theory (ISIT), 2014 IEEE International Symposium on, IEEE, 2014, pp. 756–760.

[3] A. Pierson, M. Schwager, Adaptive inter-robot trust for robust multi-robot sensor coverage, in: Robotics Research, Springer, 2016, pp. 167–183.

[4] https://commons.wikimedia.org/wiki/File:Voronoi_diagram.svg.

[5] S. Martínez, F. Bullo, J. Cortés, E. Frazzoli, On synchronous robotic networks part I: models, tasks and complexity notions, in: Decision and Control, 2005 and 2005 European Control Conference. CDC-ECC'05. 44th IEEE Conference on, IEEE, 2005, pp. 2847–2852.

[6] S. Martínez, F. Bullo, J. Cortés, E. Frazzoli, On synchronous robotic networks—Part II: time complexity of rendezvous and deployment algorithms, IEEE Transactions on Automatic Control 52 (12) (2007) 2214–2226.

[7] J.R. Peters, S. Wang, A. Surana, F. Bullo, Cloud-supported coverage control for persistent surveillance missions, Journal of Dynamic Systems, Measurement, and Control (2017).

[8] F. Ng, J. Hwu, M. Chen, X. Li, Asynchronous space-time cooperative communications in sensor and robotic networks, in: Mechatronics and Automation, 2005 IEEE International Conference, vol. 3, IEEE, 2005, pp. 1624–1629.

[9] S.M. Alamouti, A simple transmit diversity technique for wireless communications, IEEE Journal on Selected Areas in Communications 16 (8) (1998) 1451–1458.

[10] P. Flocchini, G. Prencipe, N. Santoro, P. Widmayer, Gathering of asynchronous oblivious robots with limited visibility, in: Annual Symposium on Theoretical Aspects of Computer Science, Springer, 2001, pp. 247–258.

[11] J. Lin, A.S. Morse, B.D. Anderson, The multi-agent rendezvous problem-the asynchronous case, in: Decision and Control, 2004. CDC. 43rd IEEE Conference on, vol. 2, IEEE, 2004, pp. 1926–1931.

[12] A. Pierson, L.C. Figueiredo, L.C. Pimenta, M. Schwager, Adapting to sensing and actuation variations in multi-robot coverage, The International Journal of Robotics Research 36 (3) (2017) 337–354.

Fusion of information from multiple robots

Fusion of Distributed Gaussian Process Experts (FuDGE)

The more the merrier, but whom do we trust?

Dr. Kshitij Tiwari

Contents

Highlights

- Motivation for information fusion
- Overview of decentralized–disconnected setup
- Multi-robot sensing scenario
- Various notions of fusion
- Mixture-of-Experts ensemble
- Product-of-Experts ensemble
- Notational conventions
- Novel fusion mechanism called *FuDGE*

Given the limited lifespan, no matter how eager a person is, the knowledge (s)he can acquire is theoretically limited compared to all what there is to know. Most of the decisions made are then based on this limited knowledge base. Additionally, not everyone has access to the exact same knowledge base, thus, everyone has his/her own niche area of expertise. This might lead to slight conflicts with the decisions made by others. Therefore, under the assumption that humans are rational decision makers, it would become obvious that the decisions made would be the best given the setting. This is illustrated in the following scenario: Consider three different families of martial arts, viz., taekwondo, judo, and muay thai as shown in Fig. 13.1. Each of these forms has nuances which set it apart and there has been a long standing debate about which form is better. Irrespective of such debates, consider the scenario where a competition is being organized and experts from all three forms are advising a competitor of what strike to choose next. As such, an athlete would be getting conflicting instructions given that some forms are better at kicks while others at strikes, and so on. In the ring, the fighter would not have time to process all these conflicts, so, should one form be chosen then? If so, which one? This is a hard question to answer, but the optimal strategy in such a setting is to fuse all advices and come up with a composite fighting strategy. This is what is known as *information fusion* and, having done so, the odds of winning a fight are likely to be in favor of the said competitor.

Taekwondo   **Judo**   **Muay Thai**
    (A)   (B)   (C)

Figure 13.1 Illustrating the need for information fusion via multiple martial arts forms: (A) Taekwondo which allows mostly kicks above waist (from [1]); (B) Judo which allows holds and takedowns (from [2]), and (C) Muay-thai which is a full-body combat sport that used multiple body parts to strike (from [3]).

Next, an illustrative example of the composite disconnected–decentralized deployment strategy is provided to ensure that the readers grasp the notion of this strategy. The rest of the chapter then builds on this notion and addresses the *information fusion* aspect in the multi-robot environment monitoring setting under resource constraints.

13.1   Overview of the disconnected–decentralized teams

In Chap. 8, the fully decentralized informative path planning, a.k.a. *active sensing* framework called *RC-DAS*, was discussed, which is suitable for disconnected and decentralized multi-robots teams. In Fig. 12.2 presented in Chap. 12, it appears that there are multiple nodes in the hierarchy in a decentralized setup where the decisions can be made, but there is communication involved amongst multiple levels. However, when a decentralized hierarchy is coupled with a disconnected strategy to avoid any and all communications, the situation looks slightly different. For the ease of the readers, such a setting is described in Box 13.3 with an illustrative example.

Box 13.3 Decentralized–disconnected team

A *decentralized* and *disconnected* architecture brings together the best of both deployment strategies, i.e., multiple points of decision making at zero communication overhead. This can be best explained with Fig. 13.2 which illustrates a multi-sports facility setting.

Figure 13.2 Illustrating the decentralized and disconnected scenario using a multi-sports facility example [4].

Fig. 13.2 can be interpreted as follows: Consider a group of multiple athletes, each interested in a different sport, who show up at the PQR multi-sport facility. All they need to do is to decide which sport they like, navigate to its corresponding arena, and get started. Each athlete's sport remains completely unaffected by what the others want to play or their respective levels of expertise. This obviously ignores considerations like occupancy of the arena and requirements for other players for team sports. While each athlete focuses on a different sport, what makes them a "team" is the fact that they are all professional athletes and compete at, say, international levels representing the nation globally.

This is essentially what a decentralized and disconnected team would look like. Each player/agent/robot decides what is best for them and acts on it.

In the next section, this deployment strategy is described for a multi-robot environment monitoring scenario.

13.2   Multi-robot sensing scenario

Canonical to the example presented in Box 13.3, a decentralized and disconnected multi-robot team can be setup for monitoring the target phenomena. This would encompass:

- Selecting the size of the fleet
- Equipping them with apt models (like GPs)
- Mounting apt sensors for data acquisition and efficient navigation
- Selecting self-organization and communication protocols
- Deciding on a suitable termination condition (like endurance estimation, range estimation, arbitrary budget decay, etc.) to end the respective missions

For the scope of this book and particularly this chapter, the team has been assumed to be decentralized (computationally) and disconnected (communication devoid). In doing so, for the multi-robot environment monitoring application, multiple models of the environment will be obtained, which may have slightly conflicting estimates about the underlying dynamics of the environment as shown in Fig. 13.3. This is due to the fact that every robot can only observe part of the field[1], which may not provide enough training samples to generalize the dynamics over those regions that are far away. In order to resolve such conflicts amongst local models, a novel fusion technique is discussed next to fuse all local models into one globally consistent model which can now be inferred as a representation of the overall dynamics of the environment. The research question being addressed here is:

> Given multiple models of environmental dynamics, which model should be trusted?

13.3   Various notions of fusion

The problem stated above is referring to a *many-to-one* mapping dilemma wherein each robot tries to generate a model, which it thinks is accurate, but, having obtained M models from M robots, should one or all of them be selected? If one had to be chosen, then the information acquired by the others would go in vain, but if all were retained, then the underlying environmental dynamics cannot be represented until one global model is constructed. To solve this problem, a point-wise fusion of distributed GP experts, or *FuDGE*, [5] is discussed in this chapter.

Similar works in the domain of applied machine learning use the term "*fusion*" to combine multiple sets of heterogeneous sensor data using GPs as discussed in [6–9]. In the context of multiple sensors mounted on robots, the state estimation can be done effectively by "*fusion*" of noisy information provided by various sensors using Kalman filters [10–12]. Alternatively, the term "*fusion*" in the machine learning literature is used to define an ensemble of probabilistically fused prediction estimators [13–18], which is the notion that this work will be adopting. This work can be positioned at

---

[1] Owing to resource constraints.

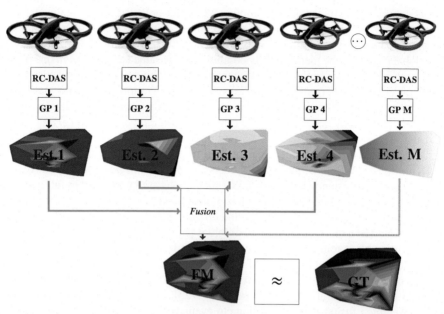

Figure 13.3 Illustration of the sensing scenario in which the team of M mobile robots operates under resource constraints. The aim is to gather optimal observations to make a prediction for the target environment. $Est.\ 1 - Est.\ M$ represent the M individualistic predictions made by each robot based on its observations, respectively. *RC-DAS* represents the active sensing scheme of choice. Fused Map (FM) is the globally consistent fused prediction map generated by merging all estimates. The objective then is to make the Fused Map as similar $(\approx)$ to the Ground Truth (GT) as possible. These maps have been interpolated for ease of visualization. In reality, these would just be a discrete collection of predicted measurements at corresponding locations. Figure based on [5].

the junction of machine learning and robotics, and the author(s) intend to use the term "*fusion*" to refer to a probabilistic amalgamation of various individually trained unbiased estimators wherein each robot itself behaves as such.

13.4 Existing fusion approaches

Existing *model fusion* techniques from the literature can be broadly classified into two main categories. The first category includes models that can be called the *Product of Expert (PoE)* models like the *Bayesian Committee Machine (BCM)* [19] and *generalized Product of Experts (gPoE)* [15]. In the BCM framework, multiple independent GP experts are trained on subsets of the whole training dataset, and their confidence is evaluated based on the reduction in uncertainty over the test points. Although this approach is promising in terms of distributing the computational load of a single GP over multiple GP experts, it is not feasible for real robot implementations. The reason for this shortcoming is quite clear: this approach works only under the assumption that all GP experts are "jointly trained" such that they "share the same set of hyperparameters" [13]. Doing so from a machine learning perspective, i.e., implementing on a workstation with sufficient computation power, is feasible and can be realized, e.g., by deploying multiple threads, each of which trains a local GP model over the subset of training data in a synchronized fashion and shares the same set of hyper-parameters. The global model can be hierarchically combined or the same can be done in one pass. On the other hand, for a real robot team, this would require precise time synchronization between all members and an all-to-all synchronized communication (as was used in the recent work [20]) in order to ensure joint training over subsets of a dataset. This problem can be easily tackled using the *generalized Product of Experts (gPoE)* models from [15], wherein the fusion is carried out over independent GP experts while their contributions are determined, e.g., by their respective differential entropy scores; cf. Definition 13.2. Both of the above models are *log opinion pool models* but BCM model ensures consistency in the sense that predictions are guaranteed to fall back to the prior when the testing data points fall significantly far away from the training data.

The second category can be called as the *Mixture of Expert (MoE)* models [16–18] wherein each GP expert specializes in different partitions of the state space and the mixture ensemble automatically allocates the expert its corresponding specialist zone. This model is a *linear opinion pool* of experts where the weights are given by input-dependent gating functions. In order to design apt gating functions, some hints can be taken from the neural network literature [21–23], which introduced a point-wise locally weighted fusion (LWF) technique to evaluate the performance of a predictor over a neighborhood around the probe point. However, these approaches require a sufficiently dense training dataset with access to ground truth. Hence, they cannot be applied directly in a real robotic setup wherein the robot never knows the ground truth. Even after visiting and observing a certain location, the robot only acquires a noisy variant of the ground truth.

Beyond the above-mentioned solutions, there are other solutions in the literature that deal with multi-agent decentralized exploration like [24] wherein a Dirichlet Process Mixture of GP (DPMGP) experts is used to model a decentralized ensemble of GP experts. In this approach, the requirement of a control parameter (α), which manages the addition of a new cluster, enforces the need of supervision (by base node or human operators) that can control and instruct a new member to be added to the team when a new cluster is created.

Aside from the above two categories of model fusion, there are other stand-alone researches. For instance, an alternative solution was proposed for multiple GP experts for decentralized data fusion by the authors in [14]. This work does not belong to any of the two categories summarized above, and, in this work, the robots share the measurements gathered with their nearest neighbors using consensus filtering [25–27].

13.5 Limitations of existing works

Robots may need to operate in harsh environments where peer-to-peer and peer-to-base communication channels are unreliable or sometimes even costly in terms of transmission costs (power consumption, latency, etc.). In such scenarios, e.g., subsurface exploration, sharing the information with the peers is infeasible. Besides, if the sensing area to be monitored is significantly large, then there is a high likelihood that the peers may never meet each other and in such settings the solutions like the decentralized data fusion schemes proposed in [14] would not suffice. In [14], consensus filtering based local communications occur amongst peers as follows: Each member of the team generates a local "summary" based on its own observations acquired. Then, when the peers are found in close proximity, the local summaries are assimilated to obtain a globally consistent "summary". This global summary of observations is then utilized to predict the mobility-on-demand requirements based on traffic conditions.

Similarly, when a multi-robot team is tasked with observing a target phenomenon of interest, members are usually not swapped, added, or removed dynamically while the team is actively exploring. This renders the work of [24] ineffective. In [24], a Dirichlet process mixture of GP experts model was proposed wherein, based on a control parameter (α), new clusters are added or removed dynamically.

As opposed to these, here an iterative weighted fusion technique suitable for GPs is described that allows evaluation of the proximity of a probe (test) point to the training samples of GP experts while evaluating the confidence of each expert. For this, first the notational conventions are re-iterated and then the model is described.

13.6 Notational conventions

Similar to the notations introduced in Chap. 8, let *dom* represent the domain of the target phenomenon within which the robot is confined to gather observations and let M represent the size of the multi-robot team. Then, $\forall m \in M$, let $U_m \subset dom$ be the set of unobserved (inputs) locations and let $O_m \subset dom$ be the set of observed (inputs) locations. Additionally, for fusion, let $U_{global} \triangleq \{U_1 \cap U_2 \cap \cdots \cap U_M\}$ represent the superset of all unobserved nodes that were never visited by any robot. Similarly, let $O_{global} \triangleq \{O_1 \cup O_2 \cup \cdots \cup O_M\}$ define the superset of all observed nodes that were visited by all robots.

13.7 Predictive model fusion for distributed GP experts (FuDGE)

At the end of the mission of all members of the mobile robot team, M diverse GP experts are acquired, which were trained on their respective subsets of training data (acquired observations) and generated a predictive map over the entire target phenomenon. To fuse the predictions from multiple models into one globally consistent model, first, a *consistency check* is performed, which involves finding the probe (test) locations that are shared by all the GP experts.

13.7.1 Fusion strategy

Let a probe point be represented by $Q \in U_{global}$ and defined as a point of interest for which the predictions from multiple GP experts must be fused. The fusion algorithm is defined next.

13.7.1.1   Point-wise mixture of experts using GMM

In what follows, first, a premise of the fusion algorithm is outlined followed by the detailed description of the fusion algorithm itself.

- **Premise:**

  GPs are kernel-based methods as explained in Chap. 6, and the isotropic squared exponential kernels as shown in Table 6.1 are used here. By definition, the correlation between locations decays exponentially as the spatial separation increases. Thus, the predictions are made with highest confidence nearby the observed locations and the confidence drops as the distance increases [28]. This is also supported by Tobler's first law of geography which states that: "Everything is related to everything else, but near things are more related than distant things" [29].

- **Model description:**

  Having laid down the premise of the model, the author(s) believe that the readers have a good intuition about the nature of correlations (given by squared exponential kernel from Table 6.1) in the environment monitoring phenomenon. Thus, now is the right time to introduce the model fusion technique hereby referred to as *FuDGE*. For this, first, independently[2] trained GP experts are obtained by utilizing the distributed GP framework from [30] and running the RC-DAS information acquisition function. Then, during the test phase, the expert predictions need to be combined based on the proximity of a test (probe) point to the experts' training samples. Thus, on the lower level, independent prediction models are deduced, and on the higher level a fused globally consistent model is obtained making this a 2-layer model.

  The length scales inferred by the GP experts represent the standard deviation in the spatial variation of measurements along the ith input dimension σ_i. A probe point Q lying too far[3] away from the training points of the mth expert will not be predicted confidently by the mth GP. This is attributed to the fact that a stationary squared exponential covariance kernel was used to model the environment. Using this covariance structure, it was inferred that the correlation between the measurements at any two locations x and x^* will decay as the spatial separation between them increases as was also explained previously in the premise of the model. Thus, a multi-variate Gaussian distribution can be placed over the O_m as $\mathcal{N}\left(Q|O_m^j, \Sigma_m\right)$ where j represents the jth training sample of the mth expert and $\Sigma_m \triangleq \text{diag}(l_{lat}^2, l_{long}^2)$.[4] The spread of the multivariate normal distribution is defined in terms of length scales along the *Latitude* and *Longitude*[5] of the corresponding GP expert. This gives rise to one Gaussian mixture

---

[2] Not the same as conditional independence. It just refers to individual models maintained by each expert.

[3] Outside the 99.5% confidence region.

[4] Not necessarily confined to latitude and longitude only. Could be any measure of spatial expanse.

[5] Used here to suit the dataset for empirical validation.

model (GMM) over the training data points of each GP expert. The log-responsibilities of this hierarchical GMM are then defined by Definition 13.1.

Definition 13.1 (Responsibility). Let x_j refer to $[O_{global}]_j$, Σ_m refer to the covariance of the Gaussian distribution for the mth GP expert and $p(Q|x_j, \Sigma_m) = \mathcal{N}(x_j, \Sigma_m)$. Then, the responsibility of the mth GP expert over the probe point $Q \in U_{global}$ is given by

$$\zeta(m|Q, O_{global}) \triangleq \sum_{x_j} \log p(Q|x_j, \Sigma_m). \tag{13.1}$$

As an intuitive illustration of the notion of *responsibility*, consider Box 13.4 which illustrates allocation of *responsibility* in case of a household fire.

Box 13.4 Responsibility

Consider the following scenario: A housing society consisting of seven houses labeled from A to G are shown in Fig. 13.4. Now, house A catches fire, perhaps due to an electrical fault in the wiring of the house. An insurance agent (observer) is sent immediately to inspect the source of fire and allocate liability to the members of the society. In other words, the agent is supposed to inspect which members will be penalized for the damages caused. For this, the agent assigns the probabilities (shown in %) to members of all seven households, and the higher the percentage, the more "responsible" the members of the said household. Intuitively, the people that are closest to the source of fire are the most liable. In other words, the closer the members of a household are to the source, the more confident is the agent they played a part in the fire. This confidence, and hence the responsibility of the members, drops as the distance from the source of fire increases.

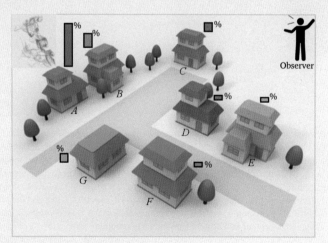

Figure 13.4 Illustration of the responsibility.

Building on this intuition, the agent assigns the highest responsibility to household A and this drops as the houses further from the source of fire are considered. For instance, houses E and F have negligible responsibility in this accident given their distance from the source of fire.

The *responsibility* in the multi-robot setting for environment monitoring is shown in Fig. 13.5 where all the locations that were not visited by robot m during its exploration are referred to as the test set for that robot, U_m. The responsibilities of a hierarchical Gaussian mixture model in Eq. (13.1) are such that $\zeta(m|Q, O_{global}) \in [0, 1]$ and $\sum_{m=1}^{M} \zeta(m|Q, O_{global}) = 1$. Then, the fused prediction at the probe point Q can be represented as the weighted fusion of predictions from all models as

$$\mu_Q \triangleq \sum_{m=1}^{M} \left(\zeta(m|Q, O_{global}) \mu_m^Q \right). \tag{13.2}$$

In Eq. (13.2), the fused prediction at probe point Q, (μ_Q), given the observations $O_{global}{}^6$ and the optimal hyper-parameters $\theta_{\forall m}$ of all the M agents is defined as the sum of predictions (μ_m^Q) weighted by the sum of the responsibilities of a GMM $\left(\zeta(m|Q, O_{global}) \right)$ for each expert $m \in \{1, \ldots, M\}$.

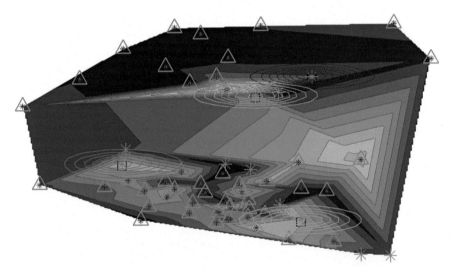

Figure 13.5 Illustration of weighted fusion performed using *FuDGE* by positioning a $2D$ Gaussian distribution $\mathcal{N}(x_i, \Sigma_m)$ to evaluate the responsibility of a GP expert over a probe point. In this figure, locations marked by a green asterisk (*) represent the training locations that were visited by the robots during their respective missions, while those highlighted by a red asterisk (*) represent the probe points over which the predictions are to be fused and black squares (□) represent the start location of each of the four robots. For ease of visualization, only the first training sample of each GP expert is shown and the process is iteratively carried out over all query points. For this illustration, four experts were considered, each of which is represented by a Gaussian contour plot centered around its first training sample [5].

---

[6] These factors have been dropped for brevity.

Additionally, the confidence of the fused estimator at the probe point Q can be explained by the net variance at the probe point Q as follows:

$$\sigma_Q \triangleq \sum_{m=1}^{M} \left\{ \zeta(m|Q, O_{global})[(\sigma_m^Q)^2 + (\mu_m^Q)^2] \right\} - (\mu_Q)^2$$

$$= \sum_{m=1}^{M} \left(\zeta(m|Q, O_{global})(\sigma_m^Q)^2 \right) + \tag{13.3}$$

$$\sum_{m=1}^{M} \left(\zeta(m|Q, O_{global})(\mu_m^Q)^2 \right) - (\mu_Q)^2.$$

Since $(\cdot)^2$ is a convex operator, using the Jensen's inequality [31], it can be concluded that $\sum_{m=1}^{M} \left(\zeta(m|Q, O_{global})(\mu_m^Q)^2 \right) \geq (\mu_Q)^2$. Now, Eq. (13.3) can be interpreted as the weighted combination of variances of the components plus a correction term which is always positive. The correction term accounts for the divergence of respective component means (μ_m^Q) from the mean of the mixture (μ_Q) for the probe point Q.

The *FuDGE* approach is summarized in Algorithm 5, and the details are as follows: The algorithm requires posterior estimates from all experts ($\forall_{m,Q} \mu_m^Q$) along

---

Algorithm 5 FuDGE ($\forall_{m,Q} \mu_m^Q$, $\forall_m \boldsymbol{\theta}_m$, U_{global}, $\forall_m O_m$).

1: **Input:**
  - $\forall_{m,Q} \mu_m^Q$: predictions from all robots for all probe points
  - $\forall_m \boldsymbol{\theta}_m$: hyper-parameters from all robots
  - $U_{global} \triangleq \{U_1 \cap U_2 \cap \cdots \cap U_M\}$: locations for fusing predictions
  - $\forall_m O_m$: observations from all robots

2: **Output:**
  - $\forall_{Q \in U_{global}} \mu^Q$: *FuDGE Predictions*
  - $\forall_{Q \in U_{global}} \sigma^Q$: *FuDGE Variances*

3:
4: **for** $\forall Q \in U_{global}$ **do**
5:     **for** each robot m **do**
6:         $\Sigma_m = \text{DIAG}.(\theta_m^2[1], \theta_m^2[2])$ ▷ Variance matrix of spatial length scales
7:         **for** each observation j **do**
8:             $\zeta_m^Q = \sum_{\forall x_j \in O_{global}} \log(p(Q|x_j, \Sigma_m))$ ▷ Compute the *Responsibility*
9:         **end for**
10:     **end for**
11:     $\zeta_m^Q \leftarrow \dfrac{\zeta_m^Q}{\sum_{\forall m \in M} \zeta_m^Q}$ ▷ Normalize
12:     $\mu_Q \leftarrow \sum_{\forall m \in M} \zeta_m^Q \mu_m^Q$ ▷ Fuse the weighted predictions from all robots
13:     $\sigma_Q \triangleq \sum_{m=1}^{M} \left\{ \zeta_m^Q)[(\sigma_m^Q)^2 + (\mu_m^Q)^2] \right\} - (\mu_Q)^2$ ▷ Fused variance
14: **end for**
15: **return** μ^Q, σ^Q

---

with their respective hyper-parameters ($\forall_m \theta_m$) generated based on corresponding observations ($\forall_m O_m$). Upon performing consistency check, the set U_{global} is obtained. Then, for each probe point $Q \in U_{global}$ (line 4), all experts are queried to obtain their learnt hyper-parameters, and the covariance matrix is generated in line 6. Similarly, the responsibility of each expert is obtained in line 8 by using Eq. (13.1). The responsibilities are then normalized to transform them into weights in line 11. Finally, a weighted summation is performed to obtain the fused prediction for the current probe point Q as shown in line 12 and fused variances in line 13. After iterating over all probe points, a list of fused predictions (μ^Q) and fused variances (σ^Q) is returned in line 15.

13.7.1.2   Generalized product-of-experts model [15]

This ensemble predicts the measurement at a test point as a weighted product of predictions from all the experts for the said test point. The gPoE model allows flexibility in the definition of weights (confidence) of each expert which are adjusted based on the importance of an expert [13]. In the original work [15], a differential entropy score was used to define the weight of the experts based on the improvement in information gain between the prior and the posterior; cf. Definition 13.2. Following this definition, this work also defines the weights (β_m) of the mth expert, and the fused predictions generated by an ensemble of M GP experts are obtained as follows:

$$\beta_m = \frac{1}{2}(\log(\sigma_{m_{**}}^2) - \log(\sigma_m^2(x^*))), \tag{13.4}$$

$$\hat{\beta}_m = \frac{\beta_m}{\sum_{m=1}^{M} \beta_m}, \tag{13.5}$$

$$\mu_{U|O,\Theta}^{gPoE} \triangleq \Sigma_{UU|O,\theta_m}^{gPoE} \sum_{m=1}^{M} \hat{\beta}_m \Sigma_{UU|O_m,\theta_m}^{-1} \mu_{U|O_m}, \tag{13.6}$$

$$(\Sigma_{UU|O,\Theta}^{gPoE})^{-1} \triangleq \sum_{m=1}^{M} \hat{\beta}_m \Sigma_{UU|O_m,\theta_m}^{-1}. \tag{13.7}$$

In Eq. (13.4), the differential entropy score is defined based on Definition 13.2 and in Eq. (13.5) the confidence weight per probe point $x^* \in U_{global}$ is evaluated by finding the differential entropy between the prior variance $\sigma_{m_{**}}^2$ and posterior variance $\sigma_m^2(x^*)$ for the probe point x^* such that $\sum_{m=1}^{M} \hat{\beta}_m = 1$. The limitation, however, is that this model is overly conservative and often over-estimates the variance. Additionally, there is no correction term in the variance to rectify over-estimation, and thus the author proposed a novel fusion technique discussed above which caters to such limitations.

Definition 13.2 (Differential entropy score). Let $\sigma_{**}^2(x)$ represent the prior variance and $\sigma_*^2(x)$ represent the posterior variance over a certain location of interest x. The differential entropy score (β) is then given by the difference in the differential entropies:

$$\beta = 0.5(\log \sigma_{**}^2(x) - \log \sigma_*^2(x)). \tag{13.8}$$

13.7.1.3 Multiple mobile sensor nodes generating single GP

Given the availability of a fusion center with sufficient processing capabilities that can fuse the models of all the robots, an obvious question then arises:

> *What happens if the multiple robots were simply considered as mobile sensor nodes, each tasked with just gathering observations, while the fusion center acquires all observations and makes a single GP directly?*

Using the mobile robot team simply as sensor nodes, gathering observations for the sink node (fusion center) instead of modeling the environmental dynamics, has the following limitations:

- Leads to an in-parallel connected topology as all agents are in direct contact with the fusion center (base station). This transforms the exploration phase itself from disconnected–decentralized to in-parallel connected–centralized architecture. As such, if the base station fails at some point, the entire team will get strangulated and the model will be completely lost.
- Acquiring data from all agents in real-time would be an additional challenge and would increase the computational time.
- Robots would be performing active sensing based on fixed hyper-parameters which cannot be updated as more data is being acquired.
- Computational complexity is again cubic in the size of the data acquired by the entire team, which is much larger than the current setting.

13.8 Map-Reduce Gaussian Process (MR-GP) framework

The active sensing architecture, along with the fusion mechanism, can all be assimilated into one sequential framework as shown in Fig. 13.6. Thus, first the *RC-DAS* objective function can be used to acquire training samples and make one model per robot. A similar procedure is repeated for all M robots, each behaving like a self-sustaining GP expert in a fully decentralized setting. This is referred to as the *Map* phase. Upon termination of the mission of all robots and successful retrieval of all M robots at the base station, a one-shot fusion of all models can be performed to obtain a globally consistent model. This stage is referred to as the *Reduce* phase, and hence, the overall framework is called the Map-Reduce Gaussian Process *(MR-GP)*. The requirement for the robots to return to the base station for the fusion procedure necessitates the need for operational range estimation (discussed in Chap. 11) and information acquisition mechanisms like *RC-DAS*[†] (discussed in Chap. 8). This not only allows models to be fused at the end, but also ensures that no information acquired by any agent goes in vain.

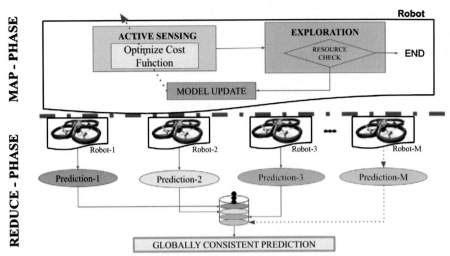

Figure 13.6 MR-GP framework showing the sequential architecture for the *Map* and *Reduce* phases. During the *Map* phase, each robot (GP-expert) generates an individual model and tries to optimize it as far as possible. Upon mission termination by all members of the team, during the *Reduce* phase, the base station performs a point-wise weighted fusion of all models to obtain a single globally consistent model. The performance of the fused model is directly influenced by the quality of each individual model.

13.9 Experiments

For empirical analysis, the **USA ozone dataset** is used in the same way as before. Based on the dataset, the model fusion quality is evaluated in unison with a variety of active sensing schemes discussed thus far. **Note:** Since NN gathers only correlated observations, the analysis with respect to nearest neighbors was omitted. Only *RC-DAS* and *full-DAS* are considered here. Also, an arbitrary fixed budget based termination criterion was used given the empirical evaluation setting. For real-world trials, this can be replaced with *xORangE*: $x \in \{online, offline\}$ range estimation framework as explained earlier in Chap. 11.

13.9.1 Fusion quality

In this section, the average RMSE is assessed, which represents the average of the errors of all robots between the estimated model and the ground truth evaluated over each element of U_{global}.

Fig. 13.7(A) shows the fusion performance of *full-DAS* against the average performance of independent robots labeled as *IndepGP*, the state-of-the-art *gPoE*, and the single GP case evaluated over U_{global}. *IndepGP* refers to the average of individual performances of all robots as evaluated over U_{global}. As explained earlier, this does not mean that the M GPs are conditionally independent of each other, since they might have had shared training samples owing to uncoordinated exploration, but independence here is used in the sense of uncoordinated individual GP expert models. It can

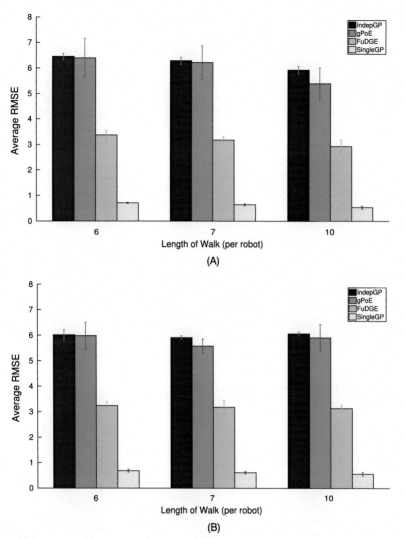

Figure 13.7 Evaluation of the average fusion performance for full-DAS (A) and RC-DAS (B) v/s Length of Walk [Ozone Dataset] [5].

be seen that the independent robots tend to incur higher (average) performance error owing to limited exploration. This error tends to go down as more observations become available. However, fusion strategies outperform the independent robot models. By comparing Figs. 13.7A and 13.7B, it can be observed that *full-DAS* tends to perform better than *RC-DAS*, since in this case the GPs had access to the most uncertain and hence, the most informative training samples. This also helps the fusion model perform better as some of the experts tend to know slightly more information about a region compared to others, and hence not all experts can be assigned equal weights. From Figs. 13.7A and 13.7B, it can be observed that the average fusion performance of the proposed model is always the best. In essence, the *FuDGE* can be considered a *Simple Averaging* when the GP experts are equally good (or bad) at predictions for a probe point, while at other times *FuDGE* assigns the weights to GP experts based on the log-likelihood (responsibility) of the GP for the probe point. Another interesting fact to note here is that, while the error of all fusion techniques for *full-DAS* tends to reduce with the increase in the number of observations (length of walk) of each robot, the error does not follow a monotonically decreasing trend for *RC-DAS* owing to the choice of training samples as explained earlier. Moreover, the *FuDGE* and *gPoE* are approximations of a single GP utilized to assist with efficient robot exploration. Thus, they incur a slight compromise in accuracy.

In order to guarantee the statistical significance of the author(s)' claims, the *p-values* [32] were calculated for the experiments. For this, consider the null hypothesis H_0: *FuDGE* does not perform better than *gPoE*, and the alternative hypothesis H_a: *FuDGE* performs better than *gPoE*. Then, for *full-DAS* and *RC-DAS*, the *p-values* of the *z-statistic* for the right-tailed test were evaluated as 0.0294 and 0.0090, respectively. The significance level of $\alpha = 0.05$ was selected and, since $p < \alpha$ for both active sensing techniques, strong evidence against the null hypothesis allows the null hypothesis to be rejected. Thus, the performance of *FuDGE* is significantly better when compared to that of *gPoE*.

13.9.2 Path length

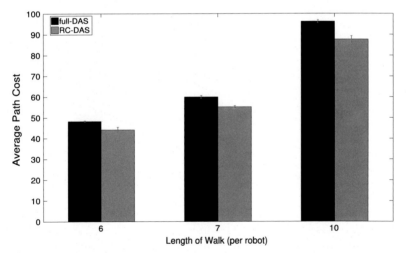

Figure 13.8 Evaluating path cost for full-DAS and RC-DAS v/s Length of Walk [Ozone Dataset] [5].

Since the active sensing schemes (informative path planning schemes discussed in Chap. 8) assist the data collection that eventually leads to map fusion, it is essential to also analyze the path cost incurred by both active sensing schemes, whereby the path cost refers to the net *sensing cost* and *traveling cost* incurred by a robot during its exploration. The results are summarized in Fig. 13.8, which shows the average past cost representing the average of total path costs incurred by all robots. It can be seen that the costs incurred by *full-DAS* are consistently higher than those of *RC-DAS*. This goes to support the claim that model performance can be successfully traded off to efficient resource utilization without having to drastically compromise on any one of them.

In conclusion, it is apparent that both *full-DAS* and *RC-DAS* attain similar performance in terms of fusion quality, but *RC-DAS* does so at lower path costs.

13.9.3 Computational complexity

From Fig. 13.7, it is evident that *FuDGE* outperforms existing state-of-the-art models, but it is also essential to realize that the performance was not obtained at the cost of extensive computations. In order to perform a fair comparison between the referred models, the trajectories of all robots were stored *a priori* using *RC-DAS* information acquisition function. Then, the same trajectories were fed to all models to simulate real robot exploration and this cost of exploration was accounted as if the model was performing exploration in real time. Let M^7 represent the size of the team operating in the field whose domain is D as before. The results hence obtained are summarized in Table 13.1, and the instance description used herewith are summarized in Table 13.2.

Table 13.1 Computational complexity analysis for FuDGE, GPoE, and SingleGP.

	Inference Complexity	Exploration Complexity	Fusion Complexity
SingleGP	$\mathcal{I}_S = \mathcal{O}(S)^3$	$\mathcal{E}_S = \mathcal{O}(R_S)$	$\mathcal{F}_S = \emptyset$
FuDGE	$\mathcal{I}_F = \mathcal{O}(\frac{S}{M})^3$	$\mathcal{E}_F = \mathcal{O}(R_F M)$	$\mathcal{F}_F = \mathcal{O}(FM)$
GPoE	$\mathcal{I}_G = \mathcal{O}(\frac{S}{M})^3$	$\mathcal{E}_G = \mathcal{O}(R_G M)$	$\mathcal{F}_G = \mathcal{O}(G + M)$

Table 13.2 Instances used for computational complexity analysis.

Instance	Description
S	$\#(O_{global})$
R_S	$\#(D \setminus O_{global})$
F	$\#(U_{global})$
R_F	$\#(D \setminus O_m)$
G	$\#(D)$
R_G	$\#(D \setminus O_m)$

In Table 13.1, the computational cost is categorized using three components: all costs referenced with $\mathcal{I}_*$ refer to the cost for performing GP inference, $\mathcal{E}_*$ refers to the computational cost for active sensing, and $\mathcal{F}_*$ refers to the computational cost for fusion. To ease the understanding of the readers, a visual representation of Table 13.1 is also shown in Fig. 13.9 wherein the model complexities are analyzed with the grow-

---

[7] The readers are hereby cautioned of a slight abuse of notation here. While in Chap. 11, the sub-script M refers to the mission, here M refers to the size of the multi-agent team. Unless stated otherwise, M should be considered to refer to size of the team.

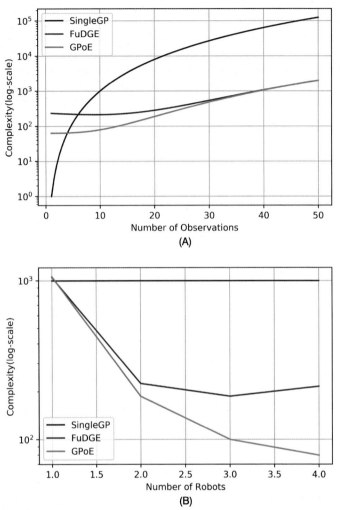

Figure 13.9 Illustration of the computational complexity of *singleGP*, *FuDGE*, and *GPoE* models: (A) Complexity v/s Length of Walk, (B) Complexity v/s Number of Robots [5].

ing number of observations for a fixed size of team (Fig. 13.9(A)) and also the impact of variable size of team (Fig. 13.9(B)). From Fig. 13.9(A), it is clear that both *FuDGE* and *GPoE* are computationally lighter than *singleGP*, and, as the number of observations grow, *FuDGE* and *GPoE* are computationally equivalent, but *FuDGE* is more accurate. In Fig. 13.9(B), it is shown that *FuDGE* and *GPoE* are better off as opposed to their *SingleGP* counterpart as they can efficiently distribute the computational load over the entire fleet. Thus, based on Fig. 13.9, the three fusion models can now be arranged in decreasing order of complexity as: $SingleGP > FuDGE \geq GPoE$. From this, it is concluded that not only *FuDGE* can generate significantly better fused maps compared to existing state-of-the-art models, but also this is done at equivalent or nominally higher computation costs. Consequently, *FuDGE* qualifies as a state-of-the-art fusion model best suited for multi-robot teams operating under resource constraints especially in the communication devoid environments.

13.10   Conclusion

This chapter introduced a novel predictive model fusion for distributed GPs called *FuDGE*. In this method, the predictions from multiple GP experts are fused point-wise while taking into account their confidence over each prediction. In doing so, globally consistent model can be obtained as negligible computational costs and maximal information assimilation. This strategy is shown to perform better over the existing state-of-the-art methods that directly aim for model fusion of multiple models, whereas this method only fuses the posteriors. This method is amicable to real-life multi-robot deployment scenarios where the heavy computations are left for the (powerful) base station which is only utilized at the end of the mission of all agents. By this time, the readers have been exposed to a variety of information-theoretic path-planning strategies, non-parametric Bayesian methods called GPs, endurance and operational-range estimation methods for real-world deployment considerations, and now with predictive fusion method for obtaining a globally consistent model to explain the environmental dynamics. In the remaining parts of the book, a brief discussion about accounting the temporal evolution of the dynamics is presented and then some success stories with real-life deployment are described before presenting conclusions and future research directions.

References

[1] https://pixabay.com/id/vectors/olahraga-seni-bela-diri-taekwondo-310088/.
[2] https://pixabay.com/id/vectors/judo-olahraga-olimpiade-logo-40769/.
[3] https://pixabay.com/id/vectors/muay-thai-seni-bela-diri-150011/.
[4] https://www.vecteezy.com/vector-art/95371-sports-vector-illustration.
[5] K. Tiwari, S. Jeong, N.Y. Chong, Point-wise fusion of distributed Gaussian process experts (FuDGE) using a fully decentralized robot team operating in communication devoid environments, IEEE Transactions on Robotics 34 (3) (2018) 820–828.
[6] S. Vasudevan, Data fusion with Gaussian processes, Robotics and Autonomous Systems 60 (12) (2012) 1528–1544.
[7] S. Vasudevan, F. Ramos, E. Nettleton, H. Durrant-Whyte, Heteroscedastic Gaussian processes for data fusion in large scale terrain modeling, in: ICRA, 2010, pp. 3452–3459.
[8] S. Vasudevan, F. Ramos, E. Nettleton, H. Durrant-Whyte, Non-stationary dependent Gaussian processes for data fusion in large-scale terrain modeling, in: ICRA, 2011, pp. 1875–1882.
[9] S. Vasudevan, A. Melkumyan, S. Scheding, Efficacy of data fusion using convolved multi-output Gaussian processes, Journal of Data Science (2015) 341–367.
[10] J. Sasiadek, P. Hartana, Sensor data fusion using Kalman filter, in: Information Fusion, 2000. FUSION 2000. Proceedings of the Third International Conference on, vol. 2, IEEE, 2000, pp. WED5–19.

[11] J. Sasiadek, Q. Wang, M. Zeremba, Fuzzy adaptive Kalman filtering for INS/GPS data fusion, in: Intelligent Control, 2000. Proceedings of the 2000 IEEE International Symposium on, IEEE, 2000, pp. 181–186.

[12] S.L. Sun, Z.L. Deng, Multi-sensor optimal information fusion Kalman filter, Automatica 40 (6) (2004) 1017–1023.

[13] M.P. Deisenroth, J.W. Ng, Distributed Gaussian processes, in: ICML, vol. 2, 2015, p. 5.

[14] J. Chen, K.H. Low, Y. Yao, P. Jaillet, Gaussian process decentralized data fusion and active sensing for spatiotemporal traffic modeling and prediction in mobility-on-demand systems, IEEE Transactions on Automation Science and Engineering 12 (3) (2015) 901–921.

[15] Y. Cao, D.J. Fleet, Generalized product of experts for automatic and principled fusion of Gaussian process predictions, in: Modern Nonparametrics 3: Automating the Learning Pipeline Workshop at NIPS, Montreal, 2014.

[16] E. Meeds, S. Osindero, An alternative infinite mixture of Gaussian process experts, Advances in Neural Information Processing Systems 18 (2006) 883.

[17] C. Yuan, C. Neubauer, Variational mixture of Gaussian process experts, in: Advances in Neural Information Processing Systems, 2009, pp. 1897–1904.

[18] S.E. Yuksel, J.N. Wilson, P.D. Gader, Twenty years of mixture of experts, IEEE Transactions on Neural Networks and Learning Systems 23 (8) (2012) 1177–1193.

[19] V. Tresp, A Bayesian committee machine, Neural Computation 12 (11) (2000) 2719–2741.

[20] A. Viseras, T. Wiedemann, C. Manss, L. Magel, Decentralized multi-agent exploration with online-learning of Gaussian process, in: ICRA, 2016, pp. 4222–4229.

[21] F. Xue, R. Subbu, P. Bonissone, Locally weighted fusion of multiple predictive models, in: IJCNN, 2006, pp. 2137–2143.

[22] P. Baraldi, A. Cammi, F. Mangili, E.E. Zio, Local fusion of an ensemble of models for the reconstruction of faulty signals, IEEE Transactions on Nuclear Science 57 (2) (2010) 793–806.

[23] F. Lavancier, P. Rochet, A general procedure to combine estimators, Computational Statistics & Data Analysis 94 (2016) 175–192.

[24] R. Ouyang, K.H. Low, J. Chen, P. Jaillet, Multi-robot active sensing of non-stationary Gaussian process-based environmental phenomena, in: AAMAS, 2014, pp. 573–580.

[25] R. Olfati-Saber, Distributed Kalman filter with embedded consensus filters, in: Proceedings of the 44th IEEE Conference on Decision and Control, IEEE, 2005, pp. 8179–8184.

[26] W. Yu, G. Chen, Z. Wang, W. Yang, Distributed consensus filtering in sensor networks, IEEE Transactions on Systems, Man and Cybernetics. Part B. Cybernetics 39 (6) (2009) 1568–1577.

[27] D.P. Spanos, R. Olfati-Saber, R.M. Murray, Dynamic consensus on mobile networks, in: IFAC World Congress, Citeseer, 2005, pp. 1–6.

[28] C.E. Rasmussen, C.K. Williams, Gaussian Processes for Machine Learning, vol. 1, MIT Press, Cambridge, 2006.

[29] N. Waters, Tobler's first law of geography, in: The International Encyclopedia of Geography, 2017.

[30] K. Tiwari, V. Honoré, S. Jeong, N.Y. Chong, M.P. Deisenroth, Resource-constrained decentralized active sensing for multi-robot systems using distributed Gaussian processes, in: 2016 16th International Conference on Control, Automation and Systems (ICCAS), 2016, pp. 13–18.

[31] M. Kuczma, An Introduction to the Theory of Functional Equations and Inequalities: Cauchy's Equation and Jensen's Inequality, Springer Science & Business Media, 2009.

[32] E.L. Lehmann, J.P. Romano, Testing Statistical Hypotheses, Springer Science & Business Media, 2006.

Part V

Continuous spatio-temporal dynamics
Continuous space-time analytics

Contents

Environmental dynamics usually vary across space and evolve over time, i.e., they are spatio-temporal in nature. Additionally, the observations are continuous in nature. Along these lines, this part summarizes the works that look into continuous spatio-temporal modeling.

V.1 Spatio-temporal analysis

This chapter primarily covers two aspects: (i) extending from discrete to continuous spatial domain, and (ii) incorporating temporal evolutions of spatial variations for full spatio-temporal analysis. State-of-the-art researches that touch upon some aspects related to these research problems are described herewith.

Spatio-temporal analysis
Temporal evolutions across continuous space

14

Time and space are not conditions of existence, time and space is a model of thinking.

Albert Einstein

Contents

Highlights

- Continuous space and time models
- Spatio-temporal kernels
- Information acquisition over continuous time

In real world, the environmental phenomena are known to vary spatially and, additionally, evolve temporally. While so far the methods focused purely on spatial modeling, this chapter addresses the design considerations when modeling complex spatio-temporal dynamics of the target phenomenon. This encompasses broadly two modifications: (i) adapting the GP kernel to account for an additional (temporal) input dimension, and (ii) adapting the *informative path planners* to accrue observations that yield insights into the complex spatio-temporal environmental dynamics. Inference under such settings is challenging and some of the state-of-the-art methods that have addressed these concerns are discussed herewith.

14.1   From discrete to continuous space

Most of the environmental monitoring data that is available via public resources like governmental environmental monitoring agencies (NOAA, etc.) are logged via static sensors. These sensors are usually spread across the region of interest. For instance, shown in Fig. 14.1 is an array of static sensors (shown in blue (●)) by NOAA monitoring ozone pollution across the continental USA. The spatial resolution achieved

Multi-Robot Exploration for Environmental Monitoring. https://doi.org/10.1016/B978-0-12-817607-8.00030-7

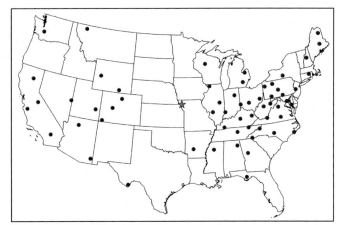

Figure 14.1 Discrete sensor placement for observing ozone concentration across the continental USA. The blue circles (●) represent the static sensor locations and the green star (★) represents the potential start location for the robot.

Ground Truth at time step - 3

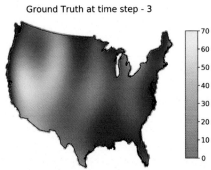

Figure 14.2 Simulated GTGP for observing ozone concentration across the continental USA. This data is now continuous across space and time. GTGP shown for $t = 3$.

by such an array of static sensors is usually very low as is evident from the figure. This leads to discretization of the state space when training GPs. Under such circumstances, if a robot were to start from the location shown by the green star (★), the obvious location for it to observe would be one of the sites marked in blue circles (●). In theory, there are no measurements recorded for locations that lie in-between the start location and the potential next location.

However, in reality, while interoceptive sensing is a valid assumption, it is usually not the case that the robot observes information at discrete and sparse points and does not get to observe anything while transiting among them. Thus, given that the GPs can *interpolate* data, it should be possible to transition from discrete to continuous state space. This was first demonstrated in [1] where the authors proposed the generation of the *Ground Truth GP (GTGP)*. In this setting, the discrete data samples from Fig. 14.1 get transformed into a continuous spatio-temporal process shown by Fig. 14.2.

The advantage of doing so is that now the robot is able to sample the data from the mean of the GP given by μ_{gt} which is continuous in space and time. The disadvantage is that the posterior (μ_{gt}) cannot be validated, owing to the lack of ground truth from static sensors at all the *interpolated* locations. Thus, GTGP needs to be trusted to be sufficiently accurate to further carry out informative and adaptive sampling. Additionally, this requires modifying the spatial kernel that was previously used in Eq. (6.1). This falls under the family of spatio-temporal kernels which were previously introduced in Chap. 5 and re-iterated here in further detail. Aside from modifying the kernel, the active sampling approach needs to be revised to account for the informativeness of samples across both space and time domains which previously focused purely on spatial domains. The following sections summarize some of the state-of-the-art works that attempt to propose the preliminary findings to such challenges.

14.2   Spatio-temporal kernels

Environmental dynamics could be varying spatially and evolving temporally across the domain. This makes modeling and inference all the more challenging as the two domains could potentially interact with each other. To this end, people have investigated spatio-temporal kernels depending on the strength of correlation between the spatial and temporal domains.

In [2], the authors proposed NOSTILL-GP-NOn-stationary Space TIme variable Latent Length scale GP, which essentially relies on a non-stationary non-separable space time kernels. As inference with a non-separable kernel is challenging (computationally), some research, like that in [3] and [4, Chap. 2], progresses by assuming separability of spatial and temporal kernel. In such cases, instead of computing the overall Gram matrix $\mathcal{K}$ directly, smaller Gram matrices are independently computed for space and time domains and then combined via a Kronecker product as follows:

$$\mathcal{K}((s,t)(s',t')) = \mathcal{K}((s,t)(s',t')) \otimes \mathcal{K}((s,t)(s',t')). \qquad (14.1)$$

In order to be able to split the Gram matrix into independent and smaller Gram matrices as shown in Eq. (14.1), a test for statistical independence must be performed *a priori*. This test is called Hilbert–Schmidt Independence Criterion (HSIC) and was introduced in [5].

14.3   Information acquisition for spatio-temporal inference

Given an apt spatio-temporal kernel, the robot now needs to be able to acquire observations that not only reveal the spatial variations but also the temporal evolution dynamics of the target phenomenon. To this end, in [1], researchers proposed continuous time path planning that learns to monitor pollution sites at locations that remain critical across multiple time steps. This, in essence, gathers training samples that can be used to train a GP that models both spatial and temporal variation scales of the polluted environment being monitored. While the acquisition function is specific to sampling from locations that exhibit high pollution levels across multiple time-scales, the robotic resource constraints are largely missing.

14.4 Summary

This chapter encompassed the adaptations in system design that need to be considered when deploying the robot teams to monitor environmental phenomena that vary spatially and evolve temporally. Some works make the assumptions that the temporal evolutions are smooth enough that they can be decoupled from spatial variations, whilst others attempt to learn both domains jointly. Simultaneous inference of both domains is a challenging task as it is, which is further complicated with the limited robot resources that restrict the mission duration severely. Whilst some research investigates optimal spatio-temporal inference mechanisms from the data scalability perspective, others attack the task from the adaptive/active path planning perspective to optimize data acquisition. While the former has seen some progress in the machine learning domain, the latter still has a vast scope of contribution specially from the robotics domain. The resource-constrained perspective that was introduced as a part of this book needs to be considered when designing such missions with complex adaptive sampling schemes. The trade-off between the *quantity* and *quality* of data that can be acquired and processed under resource limitations can largely impact the quality of models obtained, and hence must be paid heed to.

References

[1] R. Marchant, F. Ramos, Bayesian optimisation for informative continuous path planning, in: Robotics and Automation (ICRA), 2014 IEEE International Conference on, IEEE, 2014, pp. 6136–6143.
[2] S. Garg, A. Singh, F. Ramos, Learning non-stationary space-time models for environmental monitoring, in: Proceedings of the AAAI Conference on Artificial Intelligence, vol. 25, 2012, p. 45.
[3] A. Carron, M. Todescato, R. Carli, L. Schenato, G. Pillonetto, Machine learning meets Kalman filtering, in: Decision and Control (CDC), 2016 IEEE 55th Conference on, IEEE, 2016, pp. 4594–4599.
[4] S.R. Flaxman, Machine Learning in Space and Time: Spatiotemporal Learning and Inference With Gaussian Processes and Kernel Methods, Ph.D. thesis, School of Computer Science, Machine Learning Department, Carnegie Mellon University (CMU), 2015.
[5] A. Gretton, K. Fukumizu, C.H. Teo, L. Song, B. Schölkopf, A.J. Smola, A kernel statistical test of independence, in: Advances in Neural Information Processing Systems, 2008, pp. 585–592.

Part VI

Epilogue
Climax

Endings are just new beginnings.

Dr. Kshitij Tiwari

Contents

This is the final part of the book. It serves two purposes: (i) reviewing the recent successes in the real-world deployment of robots. This covers applications like algal bloom monitoring, search-and-rescue, cumulus cloud monitoring, and encompasses challenges like building accurate models with sparse data and using the received signal strength as a measure to geotag the originally unlabeled data. These aspects have been broken down into several chapters which are explained in brief below; (ii) this part also summarizes the contributions of the author(s) that were made via this book and all their earlier publications associated with this research.

VI.1   Real-world algal bloom monitoring

This chapter discusses how the robots can be utilized to monitor and curb the blooming of harmful algae which can have adverse impact of the marine ecosystem if allowed to bloom unrestricted. This encompasses dealing with challenges like the vastness of marine environments that need to be monitored along with constant variations induced by water flux, surface winds, sunlight, and other similar environment factors.

VI.2   Cumulus cloud monitoring

Yet another success story described in this part is from the cloud monitoring literature which shows how UAVs can be used together with GP models to monitor the cumulus of clouds. This is foreseen to help overcome the limitations of the existing atmospheric models by bridging the gaps between data acquired from ground and satellite measurements.

VI.3   Search & rescue

While the main focus of this book is primarily environmental monitoring applications, this chapter briefly discusses how the models presented herewith can potentially be adapted to further assist the capabilities of robots that are already being used for search-and-rescue missions. This is presented in the context of natural calamities and the search-and-rescue missions that ensue soon after. As the frequency of occurrence of the natural disasters is directly correlated to negligence towards the environment, it is only befitting to discuss the use of the presented models also in this setting to help safe lives.

VI.4   Conclusion

Being the last chapter, it presents a consolidated research summary of the author(s)' contributions described in this book and their prior works. Aside from this, potential new arenas of further research aspects are also discussed with the hope of continuing this line of research in the coming years.

Real-world algal bloom monitoring

Curbing harmful algae

Nothing is impossible. The word itself says I'm possible.

Audrey Hepburn

Contents

Highlights

- Motivation for monitoring harmful algal bloom
- Success with GPs for real-world algal bloom monitoring
- Challenges faced during such deployments
- Open research questions

This chapter discusses the attempts made at deploying GP-based environment monitoring models for real-world algal bloom monitoring using real robots. First, several marine robotic platforms are described that were designed primarily for the algal bloom monitoring application. Then, the practical challenges to the deployment of such platforms are described, followed by open research problems which hinders the widespread usability of such setups.

15.1 Motivation

Amongst the marine ecosystem dwell the countless single-celled algae called the phytoplankton. There are several varieties of these species, but amongst them are a few dozen that are harmful in the sense they either produce potent neurotoxins or cause skin irritations, illnesses, and even death in humans [1,2]. This particular subset of

Multi-Robot Exploration for Environmental Monitoring. https://doi.org/10.1016/B978-0-12-817607-8.00032-0

phytoplankton is most commonly referred to harmful algal bloom (HAB) or "red tides" and has infested virtually every coastal country in the world [2]. This is a generic term introduced by authors in [1, Chap. 1] to include all species that produce potent neurotoxins, cause water discoloration, and adversely affect the ecosystem in one form or another. Thus, to curb the adverse effects of HAB, it is necessary to understand the factors that regulate the dynamics of their blooming and then developing strategies to manage and mitigation strategies to minimize the damages.

15.2   Success with real-world deployments

In [3], a Telesupervised Adaptive Ocean Sensor Fleet (TAOSF) was presented. This fleet uses a group of Ocean–Atmosphere Sensor Integration System (OASIS) vessels developed for the National Oceanic and Atmospheric Administration (NOAA). These extended-deployment Autonomous Surface Vehicles (ASVs) enable *in-situ* study of surface and sub-surface characteristics of Harmful Algal Blooms (HAB). The robots of this fleet are shown in Fig. 15.1, and the HAB were simulated instead using a rhodamine dye (shown in red/pink in Fig. 15.1) for the demonstration of the system. The OASIS platform is a long-duration solar-powered ASV, designed for autonomous global open-ocean operations. The platform is approximately 18 feet long and weighs just over 3000 lbs [3]. The vehicle has a payload capacity of about 500 lbs, and is designed to withstand rough seas.

Figure 15.1 TAOSF fleet: overhead aerostat with camera view tethered to a manned operations boat, three OASIS platforms, and a patch of rhodamine dye. Image credits to NASA taken from [3].

Another supervised setup was considered in [4] which describes the Multilevel Autonomy Robot Telesupervision Architecture (MARTA). This is an architecture for

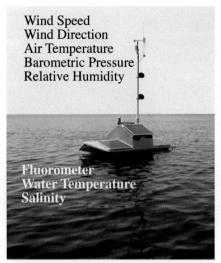

Figure 15.2 Ocean–Atmosphere Sensor Integration System (OASIS) platform with atmospheric and sea water sensors identified. Image taken from [4]. Credits to NASA.

Figure 15.3 Preliminary version of a Robot Sensor Boat (RSB). Image taken from [4] with permission from Gregg Podnar.

supervisory control of a heterogeneous fleet of networked ASVs carrying a payload of environmental science sensors that can gather salinity, conductivity, sea surface temperature, and chlorophyll measurements. This architecture allows a land-based human scientist to effectively supervise data gathering by multiple robotic assets (like those shown in Figs. 15.2 and 15.3) that implement a web of widely dispersed mobile sensors for *in-situ* study of physical, chemical, or biological processes in water or in the water/atmosphere interface. In [5], an autonomous depth-constrained underwater topography suit was presented for endowing ASVs with the capabilities to acquire bathymetric data[1] at low cost. For this research, a custom ASV was designed as shown in Fig. 15.4.

---

[1] Bathymetric data corresponds to the data representing the measurements of depth of water in water bodies like oceans, seas, or lakes.

Figure 15.4 A custom ASV developed by the Australian Center of Field Robotics (ACFR). Image taken from [5].

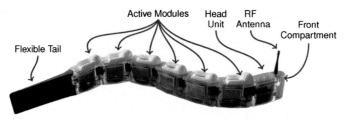

Figure 15.5 A bio-inspired environmental monitoring platform called Envirobot. Image taken from [6].

In [6], a bio-inspired snake-like marine robot for environment monitoring was presented. This robot was called Envirobot and is shown in Fig. 15.5. Some researchers like the authors of [7] have demonstrated scaling up of marine robots to a considerable fleet size (shown in Fig. 15.6) allowing for a swarm of robots to cover a wider marine environment monitoring. Das et al. [8] describe a multi-robot AUV mission planning strategies for advected bloom patches; while Podnar et al. [9] describe a multi-level telesupervision mechanism which is along the same lines as the works discussed above, whilst allowing new agents to be dynamically incorporated at different levels of the hierarchy.

15.3   Common deployment strategies

This section describes the most commonly used deployment strategies for the robots discussed above. These robots are often equipped with on-board sensors for gathering the water salinity, conductivity, sea surface temperature, and chlorophyll measurements. Some of these platforms like the OASIS are also equipped with a forward-

Figure 15.6 Showcasing a fleet of 6 out of a total of 10 custom built AUV that are relatively small and inexpensive. Image taken from [7].

looking digital imaging system for acquiring and transmitting images of atmospheric and sea state conditions to the scientists positioned at a remote base-station.

When scaling to multi-robot teams, the telemetry (e.g., GPS position, roll, pitch, yaw, battery voltage) and sensory data are either required to be communicated between each robotic platform via internet/satellite link or are transmitted back to the base station. The former allows a higher level of autonomy amongst the peers in a team who can decide for themselves how to co-ordinate the monitoring of the marine environment. For the latter case, the peers passively transmit the data to the base station and receive high-level control commands from the human operators.

15.4 Deployment challenges

Whilst marine ecosystem monitoring is an important application, there are several challenges to deploying robots teams for real-world assignments. For instance, bloom patches can be spatially large (of the order of several km) and unpredictable in their extent. To understand their ecology, water samples from the appropriate places need to be brought back to the lab for analysis. Such samples need to be taken several times to get a deeper understanding of the spatial variation and temporal evolution of the bloom. The blooming pattern is affected by a variety of factors like salinity, sunlight, temperature, and tidal variations. There are numerous temporal studies which have been conducted via a small number of sparse fixed measurement stations like [10–13], but these studies implicitly rely on the ability to extrapolate information to areas far away from the sampling locations and span only 1–2 hours of observation time with no discussion about the impact of tidal activity.

Studies like [14–19] focus on the currents and salinity properties, and their impact on the plumes blooming at the interface of salty and fresh water. These studies, however, lack the analysis of the estuarine dynamics and the spatio-temporal variations of blooming dynamics. The highly dynamic nature of tidal fluxes within estuaries is important to understand due to the effect it has on the mixing of oxygen and nutrients between fresh and salt-water. High resolution spatial and temporal sampling of this dynamic front is an ideal candidate for a moving sensing platform.

Aside from the interplay of multiple environmental factors, there is yet another challenge that is faced during real-world deployment. The challenge being that the noise levels in the measurements in a "local" region do not scale linearly to the entire target phenomenon. For instance, there might be a large amount of noise around the interface of fresh and saline water, but the region towards the interior of the estuary may be quite stable, and hence exhibit i.i.d. noise. In other words, the assumption of i.i.d. noise often breaks down in practice as input-dependent noise is observed. This gives rise to what is known as a *heteroscedastic process* as opposed a *homoscedastic process*. Heteroscedasticity was previously introduced in Eq. (6.4) in Chap. 6. As a quick recap, the homoscedastic setting would be

$$\mathcal{K}_\epsilon(x, x') = \mathcal{K}(x, x') + \delta\sigma_n^2$$
$$\Rightarrow \mathcal{K}_\epsilon = \mathcal{K} + \sigma_n^2 \mathbb{I},$$

(15.1)

where $\sigma_n \sim \mathcal{N}(0, \epsilon)$ while the heteroscedastic kernel would have $\sigma_n \sim \mathcal{N}(0, \epsilon(x))$ which makes the noise term dependent on input (x). This is then referred to as a *Heteroscedastic GP* which was originally introduced in [20]. As expected, input-dependent noise makes inference in heteroscedastic GPs a challenge especially for real-world deployments where real-time performance is a crucial concern.

15.5 Open research problems

There are a number of large physics based models for modeling the dynamics of oceans, coastal waters, and estuaries, e.g., the Regional Ocean Modeling System (ROMS) [21] and the Semi-implicit Eulerian–Lagrangian Finite Element (SELFE) model [22]. Due to the 3D numerical grids required to solve these models, the resolution is limited. As such, they will not capture the smaller scale dynamics of turbulent mixing around a front. Additionally, these are deterministic models and produce one realization of output predictions per run. This means that they cannot capture the uncertainties induced by other environmental interactions, which is a challenge for accurately modeling the algal bloom.

The solution then is to utilize mobile robots and investigate adaptive sampling strategies that take into account the phenomenon and environment dynamics simultaneously. In doing so, the robot can help with the acquisition of apt information from crucial locations that can reveal insights into algal blooming patterns across space and time domains.

References

[1] D.M. Anderson, P. Andersen, V.M. Bricelj, J.J. Cullen, J.J. Rensel, Monitoring and Management Strategies for Harmful Algal Blooms in Coastal Waters, Unesco, 2001.

[2] D.M. Anderson, Approaches to monitoring, control and management of harmful algal blooms (HABs), Ocean & Coastal Management 52 (7) (2009) 342–347.

[3] J.M. Dolan, G. Podnar, S. Stancliff, E. Lin, J. Hosler, T. Ames, J. Moisan, T. Moisan, J. Higinbotham, A. Elfes, Harmful algal bloom characterization via the telesupervised adaptive ocean sensor fleet, in: Proceedings of the 2007 NASA Science and Technology Conference (NSTC-07), College Park, MD, 2007.

[4] K.H. Low, G. Podnar, S. Stancliff, J.M. Dolan, A. Elfes, Robot boats as a mobile aquatic sensor network, in: Proc. IPSN-09 Workshop on Sensor Networks for Earth and Space Science Applications, 2009.

[5] T. Wilson, S.B. Williams, Adaptive path planning for depth-constrained bathymetric mapping with an autonomous surface vessel, Journal of Field Robotics 35 (3) (2018) 345–358.

[6] B. Bayat, A. Crespi, A. Ijspeert, Envirobot: a bio-inspired environmental monitoring platform, in: Autonomous Underwater Vehicles (AUV), 2016 IEEE/OES, IEEE, 2016, pp. 381–386.

[7] M. Duarte, J. Gomes, V. Costa, T. Rodrigues, F. Silva, V. Lobo, M.M. Marques, S.M. Oliveira, A.L. Christensen, Application of swarm robotics systems to marine environmental monitoring, in: OCEANS 2016-Shanghai, IEEE, 2016, pp. 1–8.

[8] J. Das, K. Rajany, S. Frolovy, F. Pyy, J. Ryany, D.A. Caronz, G.S. Sukhatme, Towards marine bloom trajectory prediction for AUV mission planning, in: Robotics and Automation (ICRA), 2010 IEEE International Conference on, IEEE, 2010, pp. 4784–4790.

[9] G. Podnar, J. Dolan, A. Elfes, M. Bergerman, Multi-level autonomy robot telesupervision, in: Proc. ICRA 2008 Workshop on New Vistas and Challenges in Telerobotics, 2008.

[10] H. Baumann, R.B. Wallace, T. Tagliaferri, C.J. Gobler, Large natural pH, CO2 and O2 fluctuations in a temperate tidal salt marsh on diel, seasonal, and interannual time scales, Estuaries and Coasts 38 (1) (2015) 220–231.

[11] L.C. Bruce, P.L. Cook, I. Teakle, M.R. Hipsey, Hydrodynamic controls on oxygen dynamics in a riverine salt wedge estuary, the Yarra River estuary, Australia, Hydrology and Earth System Sciences 18 (4) (2014) 1397–1411.

[12] J.H. Sharp, Estuarine oxygen dynamics: what can we learn about hypoxia from long-time records in the Delaware Estuary?, Limnology and Oceanography 55 (2) (2010) 535–548.

[13] R.M. Tyler, D.C. Brady, T.E. Targett, Temporal and spatial dynamics of diel-cycling hypoxia in estuarine tributaries, Estuaries and Coasts 32 (1) (2009) 123–145.

[14] D. Burrage, M. Heron, J. Hacker, T. Stieglitz, C. Steinberg, A. Prytz, Evolution and dynamics of tropical river plumes in the Great Barrier Reef: an integrated remote sensing and in situ study, Journal of Geophysical Research: Oceans 107 (C12) (2002).

[15] E.R. Levine, L. Goodman, J. O'Donnell, Turbulence in coastal fronts near the mouths of Block Island and Long Island Sounds, Journal of Marine Systems 78 (3) (2009) 476–488.

[16] D.G. MacDonald, L. Goodman, R.D. Hetland, Turbulent dissipation in a near-field river plume: a comparison of control volume and microstructure observations with a numerical model, Journal of Geophysical Research: Oceans 112 (C7) (2007).

[17] R.M. McCabe, B.M. Hickey, P. MacCready, Observational estimates of entrainment and vertical salt flux in the interior of a spreading river plume, Journal of Geophysical Research: Oceans 113 (C8) (2008).

[18] J. O'Donnell, G.O. Marmorino, C.L. Trump, Convergence and downwelling at a river plume front, Journal of Physical Oceanography 28 (7) (1998) 1481–1495.

[19] J. O'Donnell, S.G. Ackleson, E.R. Levine, On the spatial scales of a river plume, Journal of Geophysical Research: Oceans 113 (C4) (2008).

[20] Q.V. Le, A.J. Smola, S. Canu, Heteroscedastic Gaussian process regression, in: Proceedings of the 22nd International Conference on Machine Learning, ACM, 2005, pp. 489–496.

[21] A.M. Moore, H.G. Arango, G. Broquet, B.S. Powell, A.T. Weaver, J. Zavala-Garay, The regional ocean modeling system (ROMS) 4-dimensional variational data assimilation systems: Part I–system overview and formulation, Progress in Oceanography 91 (1) (2011) 34–49.

[22] Y. Zhang, A.M. Baptista, SELFE: a semi-implicit Eulerian–Lagrangian finite-element model for cross-scale ocean circulation, Ocean Modelling 21 (3–4) (2008) 71–96.

Cumulus cloud monitoring
Alternate aerial environment monitoring application

Clouds are like birds that just don't sleep. But what do they do with all this time they get?

Dr. Kshitij Tiwari

Contents

Highlights

- Motivation for deploying robots for monitoring clouds
- Challenges faced when deploying robots for cumulus cloud monitoring
- Open research challenges

Previously, in this book, aerial environment monitoring in terms of suspended particulate matter was discussed. As an alternate application, this chapter presents the monitoring of clouds. To some extent, both applications face similar challenges like the effect of wind on the concentration of pollutant or the cloud volume; vast expanse of area to be covered and sparse data available for modeling.

16.1 Motivation

Most of the climate General Circulation Models (GCMs) suffer from erroneous prediction of precipitation. This can be attributed partially to the uncertainties induced by clouds which are not completely understood. This in turn is governed by the weak entrainment as explained in [1]. Additionally, the erroneous predictions can also be attributed to limited complexity of the micro-physics models of the clouds themselves owing to the limits of the available computational hardware [2]. These erroneous predictions could potentially have adverse effects for the environments. For instance,

Multi-Robot Exploration for Environmental Monitoring. https://doi.org/10.1016/B978-0-12-817607-8.00033-2

dry/warm season crops like tomato, pepper, cucumber, okra, eggplant, garden egg, melon, pumpkin, sweet potato, etc., are most likely to fail if preventive measures are not taken to protect the crops from excess rainfall. This is highly contingent on apt and timely warning against expected precipitation, and, if the predictions are off by order of hours, it could lead to complete crop failures, costing the farmers dearly.

To alleviate these uncertainties, accurate model parameters and adequate measurements are required to accurately model the cloud dynamics. Doing so would require a dense spatio-temporal resolution of acquired data which is challenging for static sensors but is claimed to be achievable by a fleet of UAVs by works like [3].

16.2   Challenges

Deploying robots like UAVs to monitor cloud formations and gathering adequate data to take informed decisions are challenging tasks. Some of these are enlisted below.

1. **Sensing limitations.** Most of the robotics literature utilizes exteroceptive sensors for wide and long measuring ranges [3]. In this setting, within the sensing range, a large amount of data samples are acquired. On the other hand, most environment monitoring applications like cloud monitoring, or the ozone pollution monitoring discussed earlier, rely on point-sensing (interoceptive sensing as described in Chap. 3). As opposed to the exteroceptive setting, in this case, the sensory data is only available from the immediate position where the robot/sensor is located at a given point in time.
2. **Spatio-temporal evolution.** The environmental factors of interest (namely pressure, temperature, radiance, 3D wind, liquid water content, and aerosols) which impact the precipitation estimates vary spatially and evolve temporally. This means that the models required need high dimensional structures while the data stream is only available as a 1D time series.
3. **Impact of environmental factors.** Environmental factors, for instance, wind gusts, affect the cloud formation process and the correlated parameters of interest. Additionally, those have an impact on the energy consumed by the UAVs being used to monitor the clouds.
4. **Mission lifespan.** As explained in Chap. 10, UAVs have limited endurance.[1] For an effective mission, this entails that the endurance must be within the same range as the cumulus lifespan of the clouds.

Beset with these challenges, exploring the clouds and adaptively sampling the adequate parameters for accurate climate GCMs therefore remain a complex problem. Having said this, some progress has been made and the next section describes one of the most recent projects on cloud monitoring.

---

[1] This claim does not elude to research platforms powered by renewable energy sources as they have not been adopted for mainstream and high impact researches for environment monitoring.

16.3 Success story

One of the most notable success stories was the Skyscanner project [4] that terminated in January 2017. The project was targeted at the development of a fleet of autonomous UAVs to adaptively sample cumuli, so as to provide relevant sensory information to address long standing questions in atmospheric science pertaining to mapping of atmospheric variables at low altitudes. The objectives of this project were twofold: (i) to bring together a team of developers/designers to develop UAVs and optimize the flight control whilst jointly working with atmospheric scientists, and (ii) to help researchers that can utilize the acquired information and plug into machine learning models like GPs to map atmospheric variables based on actively acquired samples.

Figure 16.1 Mako aircraft used for the Skyscanner project. Image taken from [3].

As an outcome of this project, the researchers presented an information acquisition mechanism for a fleet of energy constrained mako aircraft with a wing span of 1.288 m and weight of 0.9 kg as shown in Fig. 16.1. The approach is claimed to generate energy optimal flight patterns that account for wind updrafts while gathering observations. However, a vast majority of the results were shown in simulations. The prediction errors achieved by GPs were significantly low, making them amicable for reliable predictions in cumulus cloud formations, but deploying such a fleet in reality still needs significant work. In the authors' own words from [3], a distributed architecture is "far-fetched" and it was recommended to make attempts to deploy a centralized fleet. A major bottleneck is the computational resources needed to embed GP regression framework on-board off-the-shelf UAVs. This was also the reason why during the span of this project, the author(s)' have also been able to validate only partial framework in real-world. Additionally, factors like collision avoidance, receding horizon path planning, sampling rate, etc., have been left out for further works.

References

[1] A.D. Del Genio, J. Wu, The role of entrainment in the diurnal cycle of continental convection, Journal of Climate 23 (10) (2010) 2722–2738.
[2] B. Stevens, S. Bony, What are climate models missing?, Science 340 (6136) (2013) 1053–1054.

[3] C. Reymann, A. Renzaglia, F. Lamraoui, M. Bronz, S. Lacroix, Adaptive sampling of cumulus clouds with UAVs, Autonomous Robots 42 (2) (2018) 491–512.

[4] S. Lacroix, G. Roberts, E. Benard, M. Bronz, F. Burnet, E. Bouhoubeiny, J.P. Condomines, C. Doll, G. Hattenberger, F. Lamraoui, et al., Fleets of enduring drones to probe atmospheric phenomena with clouds, in: EGU General Assembly Conference Abstracts, vol. 18, 2016.

Search & rescue
Enhancing the life saving operations

17

Never leave a man behind.

Warrior Ethos of the US Armed Forces

Contents

Highlights

- Conventional methods for search-and-rescue
- Modern methods utilizing teleoperated robots
- Future missions assisted by autonomous robots

Till this point in the book, applications specific to environment monitoring have been discussed along with the associated challenges, and related success stories. This chapter takes a step further by discussing one of the most pressing issues: Neglect towards the environment which is now causing frequent occurrences of the natural disasters. To assist with the search-and-rescue missions that ensue post such disasters, teleoperated robots are being deployed to reach out to places where the humans would not be able to reach otherwise. However, in order to reach the full potential of autonomous robots, this chapter briefly discusses how the machine learning approaches discussed herewith can be adapted and applied to this setting for future deployments.

17.1 Conventional search-and-rescue missions

Before the advent of search-and-rescue robots, the conventional search-and-rescue (SAR) missions were dependent primarily on the human first responders. As shown in Fig. 17.1, the first responders had to put their own lives at stake and go in, practically blind, to assist with search-and-rescue of potential survivors after a natural disaster. Recent dramatic events such as the earthquakes in Nepal and Tohoku, typhoon Haiyan, or the many floods in Europe and India have shown that local civil authorities and emergency services have difficulties in adequately managing crises. The result is that these crises often lead to major disruption of the whole local society. On top of the cost in human lives, these crises also result in adverse financial consequences, which are often extremely difficult to overcome by the affected countries.

In the event of large crises, a primordial task of the search and rescue services is the search for human survivors at the incident site. This is a complex and dangerous task, which too often leads to loss of lives among the human crisis managers themselves. The introduction of unmanned search-and-rescue (SAR) devices can offer a valuable tool to save human lives and to speed up the search-and-rescue process. Indeed, more and more robotic tools are now leaving the protected lab environment and are being deployed and integrated in the everyday life of citizens. Notable examples are automated production plants in the industry, but also the widespread use of consumer drones and the rise of autonomous cars in public space. Also in the world of search-and-rescue, these robotic tools can play a valuable role.

Of course, this does not mean that the introduction of robotic tools in the world of search-and-rescue is straightforward. On the contrary, the search-and-rescue context is extremely technology unfriendly, as robust solutions are required which can be deployed extremely quickly. Indeed, one crucial aspect must not be forgotten: the robotic tools must not have the objective to eliminate the need of human search-and-rescue workers! Instead, these robotic assets must be seen as yet another tool in the ample toolkit of human search and rescue workers in order to allow them to do their job better, faster, and safer.

Figure 17.1 Conventional search-and-rescue missions carried out primarily by human first responders by putting their own lives at risk. Image taken from [1].

17.2   Modern search-and-rescue missions

Enhancements in modern technology have gone a long way in revolutionizing not just the industries but also search-and-rescue missions' post a disaster scenario. Modern urban search-and-rescue missions are aided by highly trained sniffer dogs [2] to biomimetic snake-like robots [3] as shown in Fig. 17.2, along with a variety of other mobile robots to access areas that are otherwise unaccessible to humans. This was first demonstrated at Ground Zero [3] and continues to be the current setting for urban missions. In recent works by Professor Robin Murphy [4,5], small drones were used to get a bird's-eye view of the disaster areas which helped the first responders take informative decisions to efficiently plan their missions.

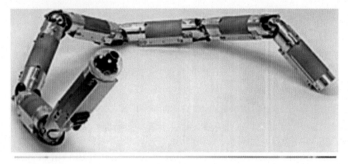

Figure 17.2 One of the many variants of the snake robot. This one was showcased in [6].

17.3   Future search-and-rescue missions

Building upon the advances in the robotics literature, researchers are now starting to optimize the system architecture in order to automate the search-and-rescue missions. In [7], an extensive survey was recently published which describes the current state of readiness and the future scope of deploying robots for post-disaster assessment. In [8], a fully autonomous aerial micro-vehicle was presented, which has the capability for navigation in both indoor and outdoor environments. Similarly in [9], autonomous payload delivery is tackled which could be useful in delivering first aid and/or essential supplies to survivors as the rescuers might need some time to safely extract all victims. A multi-robot setting was explored by [10] wherein aerial and ground robots collaborate to make a map of a building to which damage is incurred post an earthquake. In [11], a Doppler micro-radar was presented, which cancels out random body movements to measure the vital body signs without contact. The actual sensor is shown in Fig. 17.3, which is easily mountable on mobile robots given its size and configuration. However, the detection range is yet not sufficient enough to allow for long range detection or detection for humans occluded by debris, snow, obstacles, etc.

An alternate scenario could be like the one shown in Fig. 17.4. In this scenario, the aerial robot is tasked with locating human survivors and potential fires in an urban

Figure 17.3 Two identical transceivers used for random body movement cancellation. Image taken from [11].

area immediately after a natural disaster, such as an earthquake. If survivors are found in an unsafe region (denoted by red), then the robot must guide the survivors to a safe pickup region (green) before reporting its location at a data upload hub (blue). If the survivors are in a safe region or a fire is found, then the robot needs to report its location at a data upload region. Once the robot has localized survivors as well as possible, it must exit the scene (orange regions). In order to complete this task, the robot must plan its motion online by gathering information about some spatial feature (survivor locations), reacting to gain in information (by reporting locations of survivors when they are found), and organizing tasks sequentially (to ensure survivor safety before reporting locations).

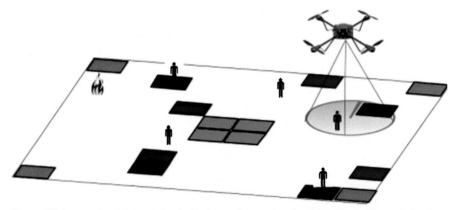

Figure 17.4 Future fire-fighting assisted with drones. Image taken from [12]. Used with permission from Austin M. Jones, Ph.D. Boston University.

17.4   Summary

While the approaches discussed above are good in their own respect, they still lack one key component: the ability to process the survivor data in order to provide the first responders with a prioritized list of areas that need to be searched. In other words, both these examples provide some information beyond just a birds-eye-view of the disaster site. This is already an enhancement over how the teleoperated robots are currently used for post-disaster analysis. But, coupling the existing state-of-the-art machine learning methods, with the advances in robotics, will help to unleash the full potential of autonomous robots and increase the chances of survival for the victims. This can partly be associated to the fact that after most disasters, it is quite challenging to locate the survivors owing to limited sensing capabilities. Of those sensors that have a larger sensing range and the capability to do so, like those presented in [13], most sensors are quite cumbersome and not mountable on off-the-shelf mobile robots. Aside from this, most machine learning models are data-driven and owing to lack of appropriate and labeled training dataset, machine learning approaches are not yet proved to be useful for such applications. The hardware and software components could be pushed beyond their current limits and potentially be combined with applied machine learning approaches. With this, it is hoped that the researchers in the coming times will further help enhance the autonomy of robots in high-stake applications like search-and-rescue.

References

[1] https://www.flickr.com/photos/alachuacounty/15588493143.

[2] W.T. Chiu, J. Arnold, Y.T. Shih, K.H. Hsiung, H.Y. Chi, C.H. Chiu, W.C. Tsai, W.C. Huang, A survey of international urban search-and-rescue teams following the Ji Ji earthquake, Disasters 26 (1) (2002) 85–94.

[3] I. Erkmen, A.M. Erkmen, F. Matsuno, R. Chatterjee, T. Kamegawa, Snake robots to the rescue!, IEEE Robotics & Automation Magazine 9 (3) (2002) 17–25.

[4] R. Murphy, J. Dufek, T. Sarmiento, G. Wilde, X. Xiao, J. Braun, L. Mullen, R. Smith, S. Allred, J. Adams, et al., Two case studies and gaps analysis of flood assessment for emergency management with small unmanned aerial systems, in: Safety, Security, and Rescue Robotics (SSRR), 2016 IEEE International Symposium on, IEEE, 2016, pp. 54–61.

[5] R.R. Murphy, B.A. Duncan, T. Collins, J. Kendrick, P. Lohman, T. Palmer, F. Sanborn, Use of a small unmanned aerial system for the SR-530 mudslide incident near Oso, Washington, Journal of Field Robotics 33 (4) (2016) 476–488.

[6] A. Wolf, H.B. Brown, R. Casciola, A. Costa, M. Schwerin, E. Shamas, H. Choset, A mobile hyper redundant mechanism for search and rescue tasks, in: Intelligent Robots and Systems, 2003 (IROS 2003). Proceedings. 2003 IEEE/RSJ International Conference on, vol. 3, IEEE, 2003, pp. 2889–2895.

[7] Jeffrey Delmerico, Stefano Mintchev, Alessandro Giusti, Boris Gromov, Kamilo Melo, Tomislav Horvat, Cesar Cadena, Marco Hutter, Auke Ijspeert, Dario Floreano, et al., The current state and future outlook of rescue robotics, Journal of Field Robotics (2019).

[8] T. Tomic, K. Schmid, P. Lutz, A. Domel, M. Kassecker, E. Mair, I.L. Grixa, F. Ruess, M. Suppa, D. Burschka, Toward a fully autonomous UAV: research platform for indoor and outdoor urban search and rescue, IEEE Robotics & Automation Magazine 19 (3) (2012) 46–56.

[9] M. Bernard, K. Kondak, I. Maza, A. Ollero, Autonomous transportation and deployment with aerial robots for search and rescue missions, Journal of Field Robotics 28 (6) (2011) 914–931.

[10] N. Michael, S. Shen, K. Mohta, V. Kumar, K. Nagatani, Y. Okada, S. Kiribayashi, K. Otake, K. Yoshida, K. Ohno, et al., Collaborative mapping of an earthquake damaged building via ground and aerial robots, in: Field and Service Robotics, Springer, 2014, pp. 33–47.

[11] C. Li, J. Lin, Random body movement cancellation in Doppler radar vital sign detection, IEEE Transactions on Microwave Theory and Techniques 56 (12) (2008) 3143–3152.

[12] https://tinyurl.com/y44r3vqz.

[13] I. Akiyama, N. Yoshizumi, A. Ohya, Y. Aoki, F. Matsuno, Search for survivors buried in rubble by rescue radar with array antennas-extraction of respiratory fluctuation, in: Safety, Security and Rescue Robotics, 2007. SSRR 2007. IEEE International Workshop on, IEEE, 2007, pp. 1–6.

Received signal strength based localization

18

Signal-strength based positioning

Unless I know where I am, I cannot possibly know if I have reached where I ought to be.

Dr. Kshitij Tiwari

Contents

Highlights

- Received Signal-Strength (RSS) based Localization
- Challenges to RSS based localization
- Scope of further research

When designing Gaussian Process (GP) regression models to fit the natural phenomenon for monitoring the spatial variations and temporal evolutions of the environment, two key factors need to be considered: (i) the measurements and (ii) the locations where these measurements are gathered from. Together, they are referred to as *inputs* to the GPs. While the on-board sensor being used for data acquisition provides the measurement of interest, what is often lacking is the location where the measurement is gathered from. This is owing to the fact that the robot operates in unknown environments and needs to "localize" itself with respect to the environment which then also informs the model of the geo-tagged observations that need to be used for training the model. This chapter describes one of the possible solutions to localization. The approach is called received signal-strength based localization (RSS), which, as the name suggests, aims at inferring the location based on the signal strength being observed by the sensors. In what follows, the state-of-the-art for such models is dis-

Multi-Robot Exploration for Environmental Monitoring. https://doi.org/10.1016/B978-0-12-817607-8.00035-6

cussed that endow coupling GPs with RSS seamlessly. This allows for one model to solve multiple problems simultaneously.

18.1   Localization based on signal strength

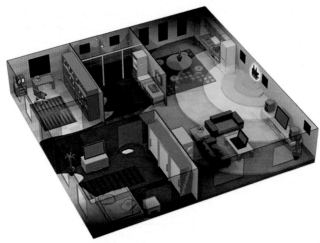

Figure 18.1 Illustration showing robot operating in an indoor environment trying to localize based on received WiFi signal strength. The closer the robot to the WiFi router, the higher the signal strength (shown in green). The farther the robot is from WiFi, the lower the signal strength, i.e., in the red zone. Used with permission from Ekahau HeatMapper www.ekahau.com.

The notion of RSS encompasses deciphering of self-location based on the strength of the measurements being acquired. Consider the illustration shown in Fig. 18.1. Given a WiFi signal-strength map in an operational environment (here indoors), the robot uses the WiFi signal strength to decode its own location. The closer the robot to the source, the higher the signal strength, and vice-versa. Mathematically, this would be represented as

$$Dist \propto \frac{1}{RSS},   \tag{18.1}$$

where $Dist$ refers to the distance from the signal source and RSS is the strength of the received signal.

In [1], Schwaighofer et al. describe a positioning system aimed at providing customized location based services to mobile users. The key idea of the Gaussian process positioning system (GPPS) is to use Gaussian process models for the signal strength received from each base station, and to obtain position estimates via maximum likelihood, i.e., by searching for the position which best fits the measured signal strengths. This work was further extended in [2], wherein the authors introduced a Bayesian filter in the architecture that can be used to represent free spaces for constraining the robot motion. While hallways, stair cases, and elevators are represented by edges in a graph, areas such as rooms are represented by bounded polygons. Using this representation, one can model both constrained motion such as moving down a hallway, or going upstairs, and less constrained motion through rooms and open spaces. The likelihood of signal strength measurements is extracted from a GP that is learned from calibration data. In contrast to existing approaches, this technique explicitly models the probability of not detecting an access point, which can greatly increase the quality of the global localization process (previously described in Chap. 4).

In [3], the authors discuss signal-strength based localization in the GPS-denied indoor environments where other signals like WiFi, which are very likely to be found in most of the indoor settings, can be used to estimate the position. In [4], the authors propose a method for RSS-based localization of robots operating in outdoor environments by proposing hardware modifications to reduce electrical interference. Additionally, a particle filter was used to fuse the odometric, RSS, and ultrasonic sensor information to bring down the localization error to sub-meter levels.

18.2   Challenges to received signal-strength based localization

There are several challenges when relying on received signal-strength based localization mechanisms. Some of them are described in further detail below along with the state-of-the-art works that serve to address these problems.

18.2.1   Lack of labeled training data

The key problem in RSS-based localization strategies is the need for labeled signal strength training data against the locations on a ground-truth map. In terms of Eq. (18.1), this refers to the fact that just by knowing the distance $Dist$ from the source, one cannot directly transform that into a 2D (or 3D) location in space unless an explicit model is provided to do so with labeled training data samples. To illustrate this, consider an example of the recent buzzword: the rabbit–duck "illusion" [5] which is shown in Fig. 18.2. Some researchers have argued against this being an illusion altogether. For instance, John F. Kihlstrom from the University of California, Berkeley, describes it as an ambiguous (bistable)[1] figure. Such optical "illusions" have confused humans for over 100 years, and now this also confuses computer vision techniques, too.[2] Thus, if both these training samples were to be fed to a machine learning model, the confidence of the model in labeling the images will constantly fluctuate between

---

[1] Further details can be found in the article here: https://www.ocf.berkeley.edu/~jfkihlstrom/JastrowDuck.htm.

[2] For an illustration, the readers are encouraged to visit https://tinyurl.com/y4cqph9k.

the image being a rabbit or a duck based on the orientation. The key takeaway point from this optical illusion and many others like this is that apt training data are a key requirement for supervised learning methods. The same is also valid for RSS-based localization settings as well.

However, this can often be prohibitive to collect or maintain as the size of the map and the domain grows. For instance, imagine labeling all the potential optical illusions explicitly in the whole wide world. This would require quite some effort. To remedy this, [6] suggest utilizing the GP-latent variable model (GP-LVM) [7] to determine the latent-space locations of unlabeled signal strength data. This model, when coupled with an appropriate motion model, can be used to reconstruct a topological connectivity graph from a signal strength sequence which, in combination with the learned Gaussian Process signal-strength model, can be used to perform efficient localization. In [8], GP-localization framework was presented that harnesses the spatial correlations in measurements to train a GP model online as opposed to relying on *a priori* acquired training data. Given that the model incurs constant time and memory overheads per filtering step, it was claimed to be optimal for persistent robot localization.

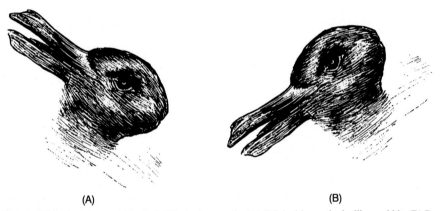

              (A)   (B)

Figure 18.2 The famous rabbit–duck "illusion" example. (A) Original image looks like a rabbit. (B) Rotated image looks like a duck. Images taken from [9].

18.2.2 Sparsity of training data

The observations that are acquired by robots or mobile sensor nodes, in general, are not only noisy but also sparse. In [10], the authors address the challenge of localization in both indoor and outdoor settings whilst acquiring sparse set of noisy observations. The basic approach is to fit an interpolant to the training data, representing the expected observation, and to assume additive sensor noise. This paper takes a Bayesian view of the problem, maintaining a posterior over interpolants rather than simply the maximum-likelihood interpolant, giving a measure of uncertainty in the map at any point.

18.2.3 Propagating location uncertainty while training

All the aforementioned works have shown meter to sub-meter accuracy in estimating the location of the robot/sensor being used to monitor the received signal strength. This induces uncertainty in localization and must be accounted for when utilizing the location information for preparing training data for GPs. There are few works that illustrate how to handle this aspect. In [11], Jadaliha et al. discuss the GP inference problem under localization uncertainty. This is an important aspect to account for as the localization mechanisms are noisy and can only provide estimates within accuracy of the order of meters.

18.3 Summary

Previous chapters in the book have described how to "map" an environmental phenomenon, and this chapter describes how to localize the robot/sensor with respect to the received signal strength of the signal being observed. Both these aspects utilize the strengths of GPs, and hence lay the groundwork for combining them to work towards simultaneous localization and mapping (SLAM). In the following chapter, concluding remarks along with an executive summary of the overall contents are presented.

References

[1] A. Schwaighofer, M. Grigoras, V. Tresp, C. Hoffmann, GPPS: a Gaussian process positioning system for cellular networks, in: Advances in Neural Information Processing Systems, 2004, pp. 579–586.

[2] B.F.D. Hähnel, D. Fox, Gaussian processes for signal strength-based location estimation, in: Proceeding of Robotics: Science and Systems, Citeseer, 2006.

[3] Y. Sun, M. Liu, M.Q.H. Meng, WiFi signal strength-based robot indoor localization, in: Information and Automation (ICIA), 2014 IEEE International Conference on, IEEE, 2014, pp. 250–256.

[4] J. Graefenstein, M.E. Bouzouraa, Robust method for outdoor localization of a mobile robot using received signal strength in low power wireless networks, in: Robotics and Automation, 2008. ICRA 2008. IEEE International Conference on, IEEE, 2008, pp. 33–38.

[5] R. Malach, I. Levy, U. Hasson, The topography of high-order human object areas, Trends in Cognitive Sciences 6 (4) (2002) 176–184.

[6] B. Ferris, D. Fox, N. Lawrence, WiFi-SLAM using Gaussian process latent variable models, in: Proceedings of the 20th International Joint Conference on Artificial Intelligence, IJCAI'07, Morgan Kaufmann Publishers Inc., San Francisco, CA, USA, 2007, pp. 2480–2485, URL http://dl.acm.org/citation.cfm?id=1625275.1625675.

[7] N.D. Lawrence, Gaussian process latent variable models for visualisation of high dimensional data, in: Advances in Neural Information Processing Systems, 2004, pp. 329–336.

[8] K.H. Low, N. Xu, J. Chen, K.K. Lim, E.B. Özgül, Generalized online sparse Gaussian processes with application to persistent mobile robot localization, in: Joint European Con-

ference on Machine Learning and Knowledge Discovery in Databases, Springer, 2014, pp. 499–503.

[9] https://tinyurl.com/c35jty.

[10] A. Brooks, A. Makarenko, B. Upcroft, Gaussian process models for indoor and outdoor sensor-centric robot localization, IEEE Transactions on Robotics 24 (6) (2008) 1341–1351.

[11] M. Jadaliha, Y. Xu, J. Choi, N.S. Johnson, W. Li, Gaussian process regression for sensor networks under localization uncertainty, IEEE Transactions on Signal Processing 61 (2) (2013) 223–237.

Conclusion & discussion
Recapitulation

19

In conclusion there is no conclusion. Things will go on as they always have, getting weirder all the time.

Robert Anton Wilson

Contents

Highlights

- Summary of author(s)' contributions
- Significance of contributions made
- Opening up arenas for further research

The aim of this book was to make an attempt to bridge the gap between the state-of-the-art machine learning models and cutting edge robots such that machine learning can make the robots fully autonomous but within the limits of current hardware. Making machine learning models work on a high performance hardware is one thing, but making the same model work on a real robot hardware is a whole new challenge. With this work, it was intended to propose models that build upon existing state-of-the-art machine learning models whilst optimizing them for real robots. The author(s)' vision is to be able to implement GPs on real robot teams and to see them observe and model real world environmental phenomena in real-time. To this end, several contributions were made as a part of this project to proceed one step closer to realizing this objective.

19.1   Summary of contributions

This work makes the following contributions:

Multi-Robot Exploration for Environmental Monitoring. https://doi.org/10.1016/B978-0-12-817607-8.00036-8

- It provides a formulation of active sensing as a bi-objective optimization with conflicting objectives to manage resources and simultaneously optimize model quality.
- It presents dynamic weight deduction for components of bi-objective optimization whilst accounting for residual resources and guaranteeing homing.
- It gives novel fusion techniques that perform point-wise fusion of predictions from various estimators weighted by their confidence over predictions. This is performed as a one-pass procedure by the base station only at the end of mission times of all robots and optimized to reduce computation by fusing predictions only over the locations which remain obscure to all models.
- It presents a novel range estimation framework generalized to encompass various classes of robots and account for various environmental conditions that a robot may be subjected to during a real mission.
- It discusses possibilities of extending the MR-GP architecture to account for temporal domain, making the architecture suited for spatio-temporal environmental modeling.

19.2   Significance of contributions

The strengths of the contributions made in this work are summarized below:

- It is rather challenging to decide the size of the team required to gather observations from the target phenomenon. To overcome this, the architecture was designed for multi-robot settings which can be easily scaled with the size of the team.
- The architecture can easily accommodate for heterogeneity in the team. This includes different nature of robots involved in a team like UAVs/UGVs, etc., along with different active sensing schemes assigned to each agent.
- The MR-GP formulation was designed for robot teams operating in communication devoid environments like sub-terrainean exploration. Thus, the work is well suited to real-world scenarios and does not bound the robots to be within communication range of the peers or the base station.
- The RC-$DAS^{\dagger}$ acquisition function was shown to be robust to starting configurations of the team. This was crucial owing to the fact that often the environment to be monitored is largely unknown and the quality of the model may be affected by the gradient of information followed from the start locations. Independence from starting configuration proves robustness of the architecture.
- The operational range estimation framework can estimate the maximum attainable range with 93% accuracy for the *online* model. This is by far the state-of-the-art range estimation framework which will prove crucial when the MR-GP is deployed on real robots and will assist RC-$DAS^{\dagger}$ to guarantee homing. Not only this, this

framework is generic enough to be utilized for any autonomous exploration mission to place upper bounds on the net path lengths that can be incurred by the robot under consideration.

- Erstwhile active sensing schemes have looked into adding resource constraints like in [1], but no prior work has tried to solve the homing problem in an information-theoretic setting.

- As opposed to other machine learning setups where the training and test dataset are pre-determined, in the spirit of active sensing, the author(s) allow the robots to choose the training and testing sets as deemed necessary to enhance the model accuracy.

- Both the active sensing and range estimation frameworks are meant to work even in the harsh conditions and handle uncertainties in the mission as far as possible.

19.3   Further works

Although the proposed framework has been validated extensively in simulations and partially for real-world scenarios, there are several ways in which the current architecture can be further enhanced. Some possibilities have been discussed below. These extensions have been further classified based on their impact on the viability of the proposed methodologies in a real-world setting.

Amongst them, some are classified as *necessary* extensions which are essential for the framework to be applied on real robots carrying out missions in real-time, as opposed to *sufficient* extensions which are simply suggested to make the framework self-sustaining without relying on external sensor information and/or human supervision. In either of these categories, the extensions are further sub-divided into *Map* and *Reduce* phases, respectively.

19.3.1   Necessary extensions

These extensions are essential for the framework to be applied on real robots carrying out missions in real-time. Such extensions have been individually discussed for the *map* and *reduce* phases.

- **Map Phase**
  - *Reducing memory footprint.* Despite adding resource constraints on active sensing, the memory cost grows as more training data is gathered. This could be solved by truncating the observations like is done in [2] wherein only a set of highly correlated training data is retained.
    For instance, consider spatio-temporal inputs of the form $\mathbf{x} \in \mathcal{R}^3 \triangleq [\mathbf{s}, \mathbf{t}]$, where, $\mathbf{s} \in \mathcal{R}^2$ and $\mathbf{t} \in \mathcal{R}$. Then, the new squared exponential spatio-temporal

covariance kernel[1] can be represented by

$$\mathcal{K}_{st}(\mathbf{x}, \mathbf{x}') = \sigma_{sig}^{2} \exp\left(-\frac{1}{2}\frac{(\mathbf{s} - \mathbf{s}')^{T}(\mathbf{s} - \mathbf{s}')}{l_{s}^{2}} - \frac{1}{2}\frac{(\mathbf{t} - \mathbf{t}')^{T}(\mathbf{t} - \mathbf{t}')}{l_{t}^{2}}\right),$$

(19.1)

where the spatial length scale is given by l_s and the temporal length scale is given by l_t. Additionally, correlations between noisy observations can be modeled by

$$\mathcal{K}_{st\epsilon}(\mathbf{x}, \mathbf{x}') = \mathcal{K}_{st}(\mathbf{x}, \mathbf{x}') + \sigma_{n}^{2}I.$$

(19.2)

The hyper-parameters are now given by $\theta \triangleq [\sigma_{sig}, l_s, l_t, \sigma_n]$. Owing to the nature of the spatio-temporal covariance kernel used in Eqs. (14.1) and (19.1), the correlation decays as the temporal separation increases. Thus, not all history is important. But, given that the team was fully disconnected and decentralized, this would also lead to loosing the correlations between observations across the agents and must be handled appropriately.

- *Location selection over receding horizon.* Currently for the *RC-DAS* and *RC-DAS*† acquisition functions, the termination of the exploration is executed if the *next-best-locations* cannot guarantee homing. But instead of doing one-shot termination, the *nth-next-best-location* could be allowed to be selected, if that guarantees homing. This could be useful since information never hurts although the *nth-next-best-location* would provide slightly correlated information where the degree of correlation depends on the magnitude of n. Therefore, even though the quality of information would be sub-optimal, it could still prove to be better than no information at all.

- *Delaying hyper-parameter update.* The active sensing schemes discussed here and those used by peers like in [1] all utilize point sensing, i.e., measurements are only obtained from the current location of the robot and the *next-best-location* (when attained). In reality, however, the sensor covers a certain region around its location and also measurements are acquired while the robot is executing the trajectories. A cumulative list of all such observations should be considered when updating the hyper-parameters of the GPs like done in [3,4].

- *Obstacle avoidance.* Obstacles (dynamic/static) have not been accounted for in the current active sensing architecture. Since the environment is largely unknown, the robot may encounter obstacles in its path as it is executing its trajectory to reach the *next-best-location*, but, owing to obstructions, it may need to replan its path.

- *Sequential optimization.* As most of the data acquired by the robots comes in sequentially, instead of inverting the whole kernel per iteration to fit the optimal hyper-parameters, sequential optimization can be utilized. As explained in

---

[1] The readers are hereby cautioned that there is a slight abuse of notation here: $\mathbf{x} - \mathbf{x}'$ does not represent subtraction between vectors. Rather it denotes pair-wise distance operation resulting in a matrix where the $(i, j)^{th}$ element represents the distance between $[\mathbf{x}_i, \mathbf{x}'_j]$.

Appendix A.1 of [5], Cholesky factors can be updated by recycling previously known factors from memory and only adding new entries corresponding to the new observations. The calculations get a bit simpler when the observations to be appended are stacked at the bottom of the observation list, i.e., new entries in Cholesky factors induce addition of new rows and columns towards the end of the matrix. This allows for a computationally efficient inference for streaming data, making the procedure amicable to real-robot setups.

- **Reduce Phase**
  - *Non-stationary heteroscedastic fusion.* In the fusion model *FuDGE*, locally stationary homoscedastic models were assumed to be sufficient, but sometimes environmental dynamics are highly non-stationary and even have heteroscedastic noise. While there are statistical tests available which can be performed *a priori* by gathering some prior measurements from static sensors to deduce the heteroscedastic nature, non-stationarity is largely the property of the model itself. In order to make the model suited to all regimes of environmental scenarios, it is advisable to fuse the locally stationary models into one globally non-stationary heteroscedastic model. Inference in such a setting would become rather challenging and hence remains open to further investigation.
  - *Active Fusion for outlier detection:* Currently, the *FuDGE* model is a passive fusion approach which relies solely on the quality of the input models being fused. In such a setting, if, during exploration, some agent went rogue owing to sensor failures or improper calculations, etc., the performance of fusion model could be affected. In order to handle such outliers, it should rather be transformed to an active fusion mechanism where the quality of local models and that of the global model are mutually affected.

19.3.2   Sufficient extensions

As opposed to the necessary extensions described above, these extensions are simply suggested to make the framework self-sustained without relying on external sensory information and/or human supervision. Just like before, these extensions are individually discussed for the *map* and *reduce* phases.

- **Map Phase**
  - *Usage of area kernels.* In [6], area kernels have be shown to be useful in merging images of high and low resolution to obtain a fine resolution fused image. This is computationally light and reflects the limitations of current hardware. However, another possible application of area kernels is active sensing. In the real world, sensors never gather point observations but rather collections of such observations (also referred to as wide and long range exteroceptive sensing). Thus, instead of handling point samples, areas/regions can be handled. Some hints on how to do inference in such a setting are available in [6]. However, the feasibility to do so in real-time is yet to be investigated.
  - *Addressing localization uncertainty.* A robot almost never knows its position with 100% certainty. Since GPs rely on location-tagged training samples which

are referred to as inputs, the uncertainty in inputs themselves must be accounted for besides the uncertainty in measurements (measurement noise) and the model uncertainty that are currently considered in the framework discussed herewith. To this end, some prior work has been done in [7].

- *GP-SLAM.* Currently, the environmental monitoring problem is posed as a *mapping* problem where the localization information is assumed to be available from external sensors. Instead, the combined mapping and localization problem, i.e., SLAM problem, can be solved with GP-LVM based Bayes filter from [8] to endow the team with co-operative localization techniques, leading to the GP-SLAM problem. SLAM or simultaneous localization and mapping [9], as known from the current literature, is used usually in terms of geometric map generation. However, in the scope of this work, GP-SLAM would be in measurement domain perhaps assisted by geometric domain to allow the robot to navigate the environment.

- *Team co-ordination for efficient exploration.* In [10], the researchers present an interesting approach for bounding the robots within their respective sensing areas by taking into account, the actuation failures and performance of each agent to shrink the respective Voronoi cell. From an information-theoretic perspective, redundancy (sensing overlaps) could prove to be useful since errors in one model could be rectified by another model. However, in the future, such team-co-ordination approaches could be investigated to see if it leads to better resource management and accurate models.

- *Non-myopic path planning.* Currently, the information acquisition functions discussed herewith are strictly *myopic*, i.e., only perform one-step look ahead. Perhaps, using some non-linear optimizers like adaptive moment estimation (ADAM) [11,12], the active sensing can be transformed into a globally optimal path planning problem.

- *Continuous time path planning.* The *RC-DAS* and its successor, *RC-DAS†*, are currently restricted only to the spatial domain. Locations observed in one time slice are considered independent of the others. However, as suggested by the authors of [13], path planning may be extended to consider the set of locations that are informative across multiple time-steps. However, using a team of multiple robots, this would involve co-ordinating the team to split the key regions amongst the team and hence will be left to future extensions. Upon successful realization, the MR-GP architecture can also be used for *forecasting* as opposed to its current application to *interpolation* as was shown in this work.

- *Anti-aliasing of observations.* Data associations have been assumed to be perfect. In reality, more than one location within the same locality or even different localities can exhibit similar measurements. Owing to such aliasing, signal strength based localization would become a challenge. Additionally, signal strength based locations are known to have an accuracy of the order to a few meters [14,15] which, the author(s) believe, can be improved by harnessing co-operative localization across the team.

- **Reduce Phase**
  - *Computational Enhancements.* As of now, the fusion architecture utilizes nested for-loops for iterating over all probe points and evaluating confidence per expert. This computational cost will increase significantly when considering the continuous spatio-temporal settings, and hence further optimization techniques need to be investigated.

- **Performance Optimization**
  - *Parallel processing of inference and interpolations.* Generating predictions over the entire spatio-temporal domain with a very high spatial resolution is usually quite slow. Currently, at every iteration of active sensing, both MLE and posterior generation are carried out sequentially. However, in order to optimize performance and distribute computational load over multiple threads, the MLE thread should run at a significantly higher frequency as compared to the posterior generation thread. The frequency of operation of these threads is a function of the data being modeled.

19.4   Closing remarks

In closing, the author(s) would like to thank the readers for expressing their interest in this project and investing their valuable time in getting to know more about the state-of-the-art. It is hoped that the book helped paint a clear picture of the current state of research when it comes to *Intelligent Environment Monitoring (IEM)* using mobile robot teams. While progress has been made, some open research questions as presented in this chapter are yet to be addressed. Thus, there will be some time and effort required before mobile robots become ubiquitous for environment monitoring and related applications. The author(s)' hope that with the contents presented herewith, not only for environment monitoring but also for alternative applications like search-and-rescue will attract growing research interest to further the research along the directions as established as a part of this project.

References

[1] R. Marchant, F. Ramos, Bayesian optimisation for intelligent environmental monitoring, in: Intelligent Robots and Systems (IROS), 2012 IEEE/RSJ International Conference on, IEEE, 2012, pp. 2242–2249.

[2] Y. Xu, J. Choi, S. Oh, Mobile sensor network navigation using Gaussian processes with truncated observations, IEEE Transactions on Robotics 27 (6) (2011) 1118–1131.

[3] H.L. Choi, J.P. How, Continuous trajectory planning of mobile sensors for informative forecasting, Automatica 46 (8) (2010) 1266–1275.

[4] W. Lu, Autonomous Sensor Path Planning and Control for Active Information Gathering, Ph.D. thesis, Duke University, 2014.

[5] M.A. Osborne, S.J. Roberts, A. Rogers, S.D. Ramchurn, N.R. Jennings, Towards real-time information processing of sensor network data using computationally efficient multi-output Gaussian processes, in: Information Processing in Sensor Networks, 2008. IPSN'08. International Conference on, IEEE, 2008, pp. 109–120.

[6] C.E. Vido, F. Ramos, From grids to continuous occupancy maps through area kernels, in: Robotics and Automation (ICRA), 2016 IEEE International Conference on, IEEE, 2016, pp. 1043–1048.

[7] S. Choi, M. Jadaliha, J. Choi, S. Oh, Distributed Gaussian process regression under localization uncertainty, Journal of Dynamic Systems, Measurement, and Control 137 (3) (2015) 031007.

[8] J. Ko, D. Fox, Learning GP-BayesFilters via Gaussian process latent variable models, Autonomous Robots 30 (1) (2011) 3–23.

[9] S. Thrun, J.J. Leonard, Simultaneous localization and mapping, in: Springer Handbook of Robotics, Springer, 2008, pp. 871–889.

[10] A. Pierson, L.C. Figueiredo, L.C. Pimenta, M. Schwager, Adapting to sensing and actuation variations in multi-robot coverage, The International Journal of Robotics Research 36 (3) (2017) 337–354.

[11] S. Ruder, An overview of gradient descent optimization algorithms, arXiv preprint, arXiv:1609.04747, 2016.

[12] D.P. Kingma, J. Ba, Adam: a method for stochastic optimization, in: International Conference on Learning Representations (ICLR), 2015.

[13] R. Marchant, F. Ramos, Bayesian optimisation for informative continuous path planning, in: Robotics and Automation (ICRA), 2014 IEEE International Conference on, IEEE, 2014, pp. 6136–6143.

[14] K.H. Low, N. Xu, J. Chen, K.K. Lim, E.B. Özgül, Generalized online sparse Gaussian processes with application to persistent mobile robot localization, in: Joint European Conference on Machine Learning and Knowledge Discovery in Databases, Springer, 2014, pp. 499–503.

[15] N. Xu, K.H. Low, J. Chen, K.K. Lim, E.B. Ozgul, GP-localize: persistent mobile robot localization using online sparse Gaussian process observation model, in: Proceedings of the Twenty-Eighth AAAI Conference on Artificial Intelligence, July 27–31, 2014, Québec City, Québec, Canada, 2014, pp. 2585–2593. URL http://www.aaai.org/ocs/index.php/AAAI/AAAI14/paper/view/8499.

List of reproduced material

Research means the activity to re-search what has already been searched. Whilst the original works may serve as motivation for further development but the original idea owners must be acknowledged.

Dr. Anupam Tiwari

A number of figures and tables in this book have been reprinted after seeking permission from the corresponding copyright owner(s). These also includes illustrations that were previously published by the author(s) at other venues and all such reprints have been duly acknowledged in the captions while the full details are mentioned below.

List of reproduced figures

Chapter 1

Fig. 1.1 reprinted with permission from https://www.pexels.com/photo/photo-of-plastics-near-trees-2583836/.

Fig. 1.2 reprinted with permission from NOAA. This figure was published on Twitter, in the article titled *Wildfire smoke from California has reached New York City, 3,000 miles away*, by National Weather Service, San Diego, Service, N.W., https://tinyurl.com/yb8ct96x. © NOAA (2018).

Fig. 1.3 reprinted with permission from https://en.wikipedia.org/wiki/2018_Sunda_Strait_tsunami#/media/File:Sunda_strait_tsunami_2.jpg.

Fig. 1.4 reprinted with permission from IEEE. This figure was published in IEEE Sensors Journal, Vol. 13 (4) in the article titled *Environmental monitoring systems: a review* by A. Kumar, H. Kim and G.P. Hancke, pp. 1329–1339. © IEEE (2013).

Fig. 1.5 reprinted with permission from liquid-robotics.

Chapter 2

Fig. 2.1 used with permission from Scentroid (http://scentroid.com/scentroid-dr1000/).

Fig. 2.2 reprinted with permission from NASA Earth Observatory.

Fig. 2.3 reprinted with permission from Deepfield Robotics. This figure was published in Intelligent Robots and Systems (IROS), 2015 IEEE/RSJ International Conference on, in the article *Vision-based high-speed manipulation for robotic ultra-precise weed control* by A. Michaels, S. Haug and A. Albert, pp. 5498–5505. © Deepfield Robotics (2015).

Chapter 3

Fig. 3.2 used with permission from Extreme Tech (https://commons.wikimedia.org/wiki/File:Quadrotor.jpg).

Fig. 3.4 reprinted with permission from Marine Geology. This figure was published in Marine Geology, Vol. 352 in the article titled *Autonomous underwater vehicles (AUVs): their past, present and future contributions to the advancement of marine geoscience* by R.B. Wynn et al., pp. 451–468. © Marine Geology (2014).

Fig. 3.5(A) reprinted with permission from https://tinyurl.com/y32xgur7.

Fig. 3.5(B) reprinted with permission from https://commons.wikimedia.org/wiki/File:KUKA_youBot.jpg.

Chapter 4

Fig. 4.1 reprinted with permission from Springer. This figure was published in Autonomous Robots, in the article titled *OctoMap: an efficient probabilistic 3D mapping framework based on octrees* by A. Hornung et al., doi:10.1007/s10514-012-9321-0. © Springer (2013).

Fig. 4.2 reprinted with permission from IEEE. This figure was published in IEEE Transactions on Robotics, Vol. 28 (3) in the article titled *On cooperative patrolling: optimal trajectories, complexity analysis, and approximation algorithms* by F. Pasqualetti, A. Franchi and F. Bullo, pp. 592–606. © IEEE (2012).

Fig. 4.3 reprinted with permission from Elsevier. This figure was published in Robotics and Autonomous Systems, Vol. 66 in the article titled *Semantic mapping for mobile robotics tasks: a survey* by I. Kostavelis and A. Gasteratos, pp. 86–103. © Elsevier (2015).

Fig. 4.4 reprinted with permission from IEEE. This figure was published in IEEE Transactions on Robotics, Vol. 34 (4) in the article titled *Modeling and interpolation of the ambient magnetic field by Gaussian processes* by A. Solin et al., pp. 1112–1127. © IEEE (2018).

Chapter 7

Fig. 7.4 reprinted with permission from https://www.pexels.com/photo/statue-room-2574476/.

Fig. 7.5 reprinted with permission from IEEE. This figure was published in Intelligent Robots and Systems (IROS), 2015 IEEE/RSJ International Conference on, in the article *Visibility-based persistent monitoring with robot teams* by P. Tokekar and V. Kumar, pp. 3387–3394. © IEEE (2015).

Fig. 7.8 reprinted with permission from SAGE. This figure was published in The International Journal of Robotics Research, Vol. 36 (3) in the article titled *Adapting to sensing and actuation variations in multi-robot coverage* by A. Pierson et al., pp. 337–354. © SAGE (2017).

Fig. 7.9 reprinted with permission from ASME. This figure was published in Journal of Dynamic Systems, Measurement, and Control, in the article titled *Cloud-supported coverage control for persistent surveillance missions* by J.R. Peters et al. © ASME (2017).

Chapter 8

Fig. 8.4 reprinted with permission from IEEE. This figure was published in 2016 16th International Conference on Control, Automation and Systems (ICCAS), in the article *Resource-constrained decentralized active sensing for multi-robot systems using distributed Gaussian processes* by K. Tiwari et al., pp. 13–18. © IEEE (2016).

Fig. 8.5 reprinted with permission from IEEE. This figure was published in Industrial Electronics Society, IECON 2017-43rd Annual Conference of the IEEE, in the article *Multi-UAV resource constrained online monitoring of large-scale spatio-temporal environment with homing guarantee* by K. Tiwari, S. Jeong and N.Y. Chong, pp. 5893–5900. © IEEE (2017).

Chapter 10

Fig. 10.3 reprinted with permission from https://commons.wikimedia.org/w/index.php?curid=29811585.

Chapter 11

Figs. 11.2, 11.5–11.8, 11.13, 11.17–11.19 reprinted with permission from IEEE. These figures were published in 2018 IEEE/RSJ International Conference on Intelligent Robots and Systems (IROS), in the article titled *Estimating achievable range of ground robots operating on single battery discharge for operational efficacy amelioration* by K. Tiwari, X. Xiao and N.Y. Chong, pp. 3991–3998. © IEEE (2018).

Fig. 11.4 reprinted with permission from IEEE. This figure was published in Robots in the Wild Workshop: Challenges in Deploying Robust Autonomy for Robotic Exploration, in the article titled *ORangE: Operational Range Estimation for mobile robot exploration on a single discharge cycle* by K. Tiwari et al. © IEEE (2019).

Fig. 11.9(A) reprinted with permission from IEEE. This figure was published in IEEE Transactions on Robotics, Vol. 31 (5) in the article titled *ORB-SLAM: a versatile and accurate monocular SLAM system* by R. Mur-Artal, J.M.M. Montiel and J.D. Tardós, pp. 1147–1163. © IEEE (2015).

Figs. 11.10–11.12, 11.14, 11.15, 11.20–11.27 reprinted with permission from Elsevier. These figures were published in Mechatronics, Vol. 57 in the article titled *A unified framework for operational range estimation of mobile robots operating on a single discharge to avoid complete immobilization* by K. Tiwari et al., pp. 173–187. © Elsevier (2019).

Chapter 12

Fig. 12.4 reprinted with permission from https://commons.wikimedia.org/wiki/File: Voronoi_diagram.svg.

Chapter 13

Fig. 13.1(A) reprinted with permission from https://pixabay.com/id/vectors/olahraga-seni-bela-diri-taekwondo-310088/.

Fig. 13.1(B) reprinted with permission from https://pixabay.com/id/vectors/judo-olahraga-olimpiade-logo-40769/.

Fig. 13.1(C) reprinted with permission from https://pixabay.com/id/vectors/muay-thai-seni-bela-diri-150011/.

Fig. 13.2 reprinted with permission from https://www.vecteezy.com/vector-art/95371-sports-vector-illustration.

Figs. 13.3, 13.5, 13.7–13.9 reprinted with permission from IEEE. These figures were published in IEEE Transactions on Robotics, Vol. 34 (3) in the article titled *Point-wise fusion of distributed Gaussian process experts (FuDGE) using a fully decentralized robot team operating in communication devoid environments* by K. Tiwari, S. Jeong and N.Y. Chong, pp. 820–828. © IEEE (2018).

Chapter 15

Fig. 15.1 reprinted with permission from NASA. This figure was published in Proceedings of the 2007 NASA Science and Technology Conference (NSTC-07), in the article titled *Harmful algal bloom characterization via the telesupervised adaptive ocean sensor fleet* by J.M. Dolan et al. © NASA (2007).

Figs. 15.2, 15.3 reprinted with permission from ESSA. These figures were published in Proc. IPSN-09 Workshop on Sensor Networks for Earth and Space Science Applications, in the article titled *Robot boats as a mobile aquatic sensor network* by K.H. Low et al. © ESSA (2009).

Fig. 15.4 reprinted with permission from Wiley. This figure was published in Journal of Field Robotics, Vol. 35 (3) in the article titled *Adaptive path planning for depth-*

constrained bathymetric mapping with an autonomous surface vessel by T. Wilson and S.B. Williams, pp. 345–358. © Wiley (2018).

Fig. 15.5 reprinted with permission from IEEE. This figure was published in Autonomous Underwater Vehicles (AUV), 2016 IEEE/OES, in the article titled *Envirobot: a bio-inspired environmental monitoring platform* by B. Bayat, A. Crespi and A. Ijspeert, pp. 381–386. © IEEE (2016).

Fig. 15.6 reprinted with permission from IEEE. This figure was published in OCEANS 2016-Shanghai, in the article titled *Application of swarm robotics systems to marine environmental monitoring* by M. Duarte et al., pp. 1–8. © IEEE (2016).

Chapter 16

Fig. 16.1 reprinted with permission from Springer. This figure was published in Autonomous Robots, Vol. 42 (2) in the article titled *Adaptive sampling of cumulus clouds with UAVs* by C. Reymann et al., pp. 491–512. © Springer (2018).

Chapter 17

Fig. 17.1 reprinted with permission from https://www.flickr.com/photos/alachuacounty/15588493143.

Fig. 17.2 reprinted with permission from IEEE. This figure was published in Proceedings 2003 IEEE/RSJ International Conference on Intelligent Robots and Systems, Vol. 3, in the article titled *A mobile hyper redundant mechanism for search and rescue tasks* by A. Wolf et al., pp. 2889–2895. © IEEE (2003).

Fig. 17.3 reprinted with permission from IEEE. This figure was published in IEEE Transactions on Microwave Theory and Techniques, Vol. 56, in the article titled *Random body movement cancellation in Doppler radar vital sign detection* by C. Li and J. Lin, pp. 3143–3152. © IEEE (2008).

Fig. 17.4 reprinted with permission from Jones, Austin, https://tinyurl.com/y44r3vqz.

Chapter 18

Fig. 18.1 reprinted with permission from Ekahau HeatMapper, www.ekahau.com.

Fig. 18.2 reprinted with permission from https://tinyurl.com/c35jty.

List of reproduced tables

Chapter 8

Table 8.1 reprinted with permission from IEEE.

Table 8.2 reprinted with permission from IEEE.

Chapter 11

Table 11.2 reprinted with permission from IEEE.

Chapter 13

Table 13.1 reprinted with permission from IEEE.

Table 13.2 reprinted with permission from IEEE.

Index